Fundamentals of Immunogenetics: Principles & Practices

SURAKSHA AGRAWAL
Head
Deptt. of Genetics
Sanjay Gandhi Post Graduate Institute of Medical Sciences,
Raebareli Road, Lucknow - 226 014 (UP), India

SITA NAIK
Head
Deptt. of Immunology
Sanjay Gandhi Post Graduate Institute of Medical Sciences,
Raebareli Road, Lucknow - 226 014 (UP), India

IBDC

International Book Distributing Co.
(Publishing Division)

Published by

INTERNATIONAL BOOK DISTRIBUTING CO.
(Publishing Division)
Khushnuma Complex Basement
7, Meerabai Marg (Behind Jawahar Bhawan)
Lucknow 226 001 U.P. (INDIA)
Tel. : 91-522-2209542, 2209543, 2209544, 2209545
Fax : 0522-4045308
E-Mail : ibdco@airtelbroadband.in

First Edition 2007

MAXIMUM RETAIL PRICE INCLUSIVE OF ALL TAXES Rs 350.00

Price:
ISBN 81-8189-153-8 PB

Composed & Designed at :
Panacea Computers
3rd Floor, Agrawal Sabha Bhawan
Subhash Mohal, Sadar Cantt. Lucknow-226 002
Phone : 0522-2483312, 9335927082
E-mail : prasgupt@rediffmail.com

Printed at:
Salasar Imaging Systems
C-7/5, Lawrence Road Industrial Area
Delhi - 110 035
Tel. : 011-27185653, 9810064311

Preface

This book is an introductory textbook of immunogenetics, an exciting and rapidly advancing branch of biology. The book evolved from a course we regularly teach to our first year Ph.D. students. Although we get bright MSc. students from many major Indian universities, the lack of exposure to immunogenetics at the post-graduate level has been striking. We felt that a simple primer on basic immunogenetics for postgraduates in biological science and medicine would be useful and fill a gap in the existing spectrum of Indian textbooks.

There have been many exciting developments in immunogenetics in the last two decades which have made major impact on human health. Students need to be conversant with these advances in order to understand the immunologic diseases and to use the rapidly emerging methods of diagnosis and therapy that are based on immunologic approaches.

This book tries to explain the basic principles of immunogenetics in a simple accurate and up-to-date manner. We have also tried to convey some of the excitement of this field and provide the foundation required for young scientists to launch a career in this field. We hope that it will help them to understand the genetic basis of diseases and generate enthusiasm for the speciality.

We would like to acknowledge the help and support we have received from many individuals. Foremost among these were colleagues who provided invaluable constructive criticisms and support. Mr. Sanjay Kumar Johari has been responsible for bringing out an excellent manuscript.

Lastly, we thank many people who have faithfully supported us even when we were not available for them - the members of our laboratories who kept our research projects alive even when we were busy in writing this book and our families, who were tolerant of our many demands and awaited for the completion of the book with an eagerness that matched our own.

Contents

CHAPTER 1

Introduction to Immune System

Living organisms provide the ideal habitat for other organisms to grow. It is not surprising, therefore, that animals are subject to infections by viruses, bacteria, fungi and parasites. The immune system has evolved in such a way that it provides our body with the necessary mechanisms to fight against a variety of organisms, and maintain a continuous surveillance against disease. The strategies employed in this battle are highly sophisticated in higher vertebrates. It is a combined activity of many different cells, some of which patrol the body, while others are concentrated in various lymphoid organs. These strategies are developed in a way that can selectively destroy foreign molecules and cells without harming the host's own normal cells. The weapons of the immune system include (i) cells that kill or ingest infected or altered cells and (ii) soluble proteins that can neutralize, immobilize, agglutinate or kill pathogens. The immune system is also implicated in the body's fight against cancer but the degree to which the system can identify and kill cancer cells remains controversial. This ability to eradicate infectious agents or cancer cells is termed immunity. The immune system retains a memory of each attack, allowing it to respond more efficiently to subsequent attacks by the same invader. This is the reason why individuals rarely suffer more than once from the same infectious disease such as chickenpox, mumps, or whooping cough. Immune mechanisms become activated when the body encounters various pathogens and failure to mount protective immune response or an aberrant and excessive response cause disease.

There are two basic strategies the body can use to protect itself against infections on the basis of which the immune system is divided into adaptive immune response and non-adaptive immune response. While the latter is an immediate response to any foreign agent, adaptive immune response occurs against pathogens and confers long-standing protection. It is more specific in nature as compared to non-adaptive immunity. The organs involved in the immune response are the thymus, spleen, lymph nodes and cells of haematopoietic origin distributed throughout the body **(Figure 1.1)**

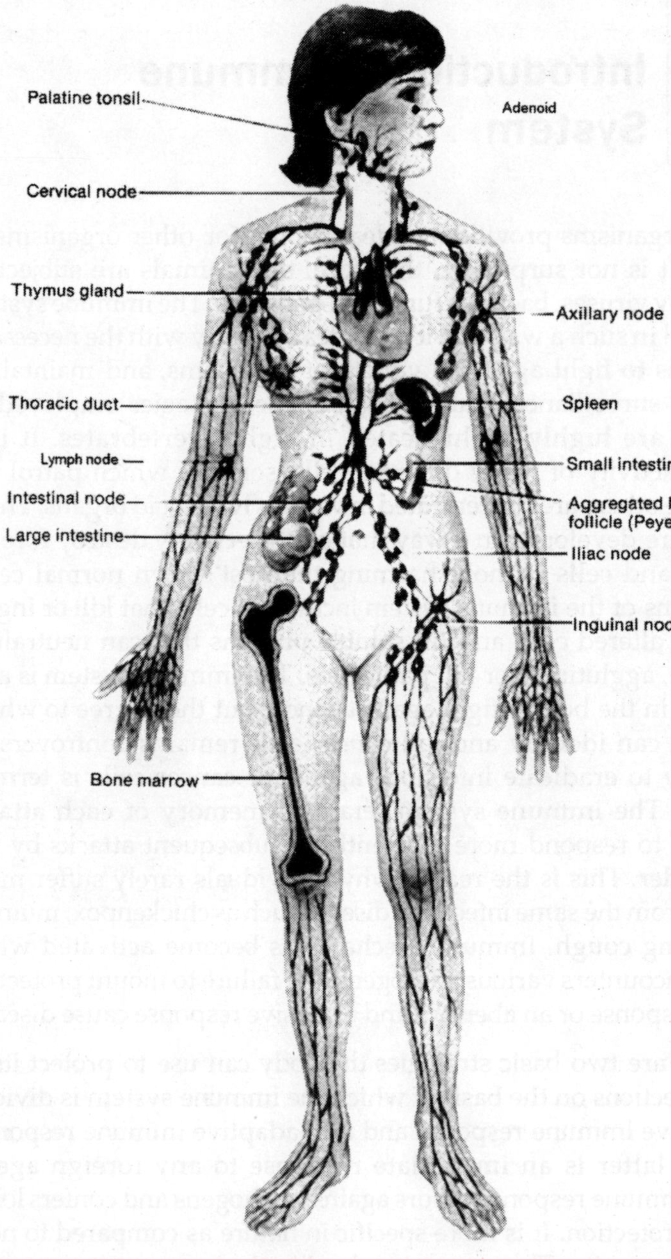

***Figure 1.1 The human immune system** includes various lymphoid organs, such as the thymus, bone marrow, spleen, lymph nodes and scattered cells located in patches within the small intestine, adenoids, and tonsils.*

The earliest demonstration of immunological principles was the successful vaccination against smallpox developed by Edward Jenner in England in the late 18th century. At that time he knew little about either the pathogen or the body's response to it. The studies of Koch and Pastuer in the 19th century began to unravel the science of "Immunology".

The immune response may be divided into two broad overlapping parts based either on time kinetics of the response or on functional events. The early response or first line of defence occurs within minutes to hours of encounter with foreign antigen and is termed innate or non-specific immunity or non-adaptive immunity.

Innate immune responses

Innate responses are characterized by their lack of specificity and the same set of defensive activities is harnessed regardless of the pathogens presenting the threat. Innate responses to invading bacteria are typically accompanied by a process of inflammation at the site of infection where cells and dissolved substances leak out of the blood into the affected tissues. These events are accompanied by local redness, swelling and fever. Inflammation provides a means for concentrating the body's defensive agents at the site where they are needed. Phagocytic cells such as neutrophils and macrophages leave the bloodstream and migrate towards the site of infection in response to chemo attractants (chemicals) released at the site. These cells recognize, engulf and destroy the pathogens (**Figure 1.2a &1.2b**). Innate immune responses are particularly effective against bacteria whose cell walls contain unique polysaccharides. Infected cells are also eliminated by the process of apoptosis through natural killer (NK) cells (**Figure 1.2c**). The innate immune response against viral infections is predominantly mediated by interferon alpha (**Figure 1.2d**). The delayed response occurs within days and is known as adaptive or specific immunity. The recognition events are termed as the afferent arm and the effector functions are known as efferent arm. In the efferent response effector functions may be mediated either through cells (cell mediated immunity) or antibodies (humoral immunity).

The unique features of the immune response, which distinguish it from all other physiological responses of the body, are specificity, diversity and memory. The response is specific not to each antigen but to the structural components of complex protein, polysaccharide

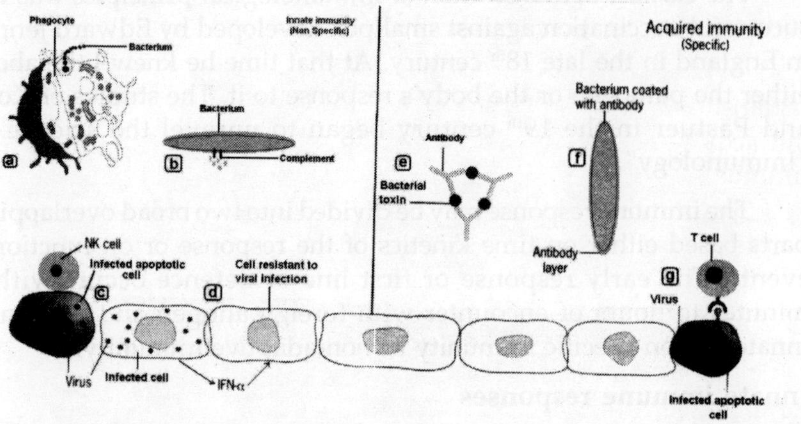

***Figure 1.2 An overview of some of the mechanisms by with the immune system rids the body of invading pathogens.** The left panel depicts several types of innate immunity: (a) phagocytosis of a bacterial cell (b) bacterial cell killing by complement (c) apoptosis of an infected cell by a natural killer (NK) cell (d) induction of viral resistance by Interferon-α. The right panel depicts several types of acquired immunity (e) neutralization of a bacterial toxin by antibody (f) bacterial cell coated with antibody (opsonized), which makes it susceptible to phagocytosis or complement induced death (g) apoptosis induced in an infected cell by an activated T lymphocyte (T cell).*

and other antigens; these are termed epitopes or determinants. Specificity is conferred by the unique receptors on the T and B-lymphocytes responding to the antigens, which recognize even subtle differences between antigens.

Diversity refers to the capability of the immune system to respond to an extremely wide range of antigen specificities.

Memory is the imprint of itself that an antigen leaves behind after a successful immune response is generated. This retention of the imprint allows the body to respond more rapidly to subsequent contact with the same antigen.

Acquired immune responses

Unlike innate responses, acquired immune responses require a lag period during which the immune system gears up for an attack against a foreign agent. Its characteristic feature is its high specificity that is, they are directed only against specific molecules present in the foreign material. For example, the blood of a person who has just recovered from measles contains antibodies that react with the virus that causes measles but not with a related virus, such as that which causes mumps.

Whereas all animals possess some type of innate immunity against microbes and parasites, only vertebrates are known to mount an acquired response. There are two broad categories of acquired immunity.

- **Humoral immunity** is mediated by antibodies, which are globular, blood-borne proteins of the immunoglobulin (Ig) super family
- **Cell-mediated immunity** is mediated by cells.

Both types of acquired immunity are mediated by lymphocytes, which are nucleated leukocytes (white blood cells) that circulate between the blood and lymphoid organs. Humoral immunity is mediated by B lymphocytes (or B cells), when activated, differentiate into cells which secrete antibodies. Antibodies are directed primarily against foreign materials such as protein and polysaccharide components of bacterial cell walls, bacterial toxins and viral coat proteins. In some cases, antibodies can bind to a bacterial toxin or viral coat protein and directly prevent the agent from entering a host cell (**Figure 1.2 e**). In other cases, antibodies function as "molecular tags", bind to an invading pathogen and mark it for destruction. Bacterial cells coated with antibody molecules are rapidly ingested by wandering phagocytes or destroyed by complement molecules carried in the blood **(Figure 1.2f).** Antibodies are not effective against intracellular pathogens, which are dealt with by a second type of weapon system or cell-mediated immunity. This is mediated by T lymphocytes (or T cells), when activated, can specifically recognize and kill an infected (or foreign) cell (**Figure 1.2 g**).

B and T cells arise from the same type of precursor cell (*a pluripotent haematopoietic stem cell*), but differentiate along different pathways in different lymphoid organs. A summary of the various pathways of differentiation of the *pluripotent haematopoietic* stem cell is shown in **Figure 1.3**. B-lymphocytes differentiate in the fetal liver or adult bone marrow, whereas T lymphocytes differentiate in the thymus gland, an organ located in the chest that reaches its peak size during childhood. Because of these differences, cell-mediated and humoral immunity can be dissociated to a large extent. For instance, humans may suffer from a rare disease called congenital agammaglobulinemia in which humoral antibody is deficient and cell mediated immunity is normal.

The clonal selection theory as it applies to B cells

In case of infection with any organism, the blood soon contains a high concentration of antibodies capable of reacting with the foreign substance. These antibodies are formed to the various components of infecting agents, which are termed 'antigens'. Most antigens are proteins or polysaccharides, but lipids and nucleic acids can also be antigens. How can the body produce antibodies that react specifically with an antigen to which the body is exposed? In other words, how does an antigen induce an acquired immune response? For many years it was thought that antigens somehow instruct lymphocytes to produce complementary antibodies. It was suggested that an antigen wraps itself around an antibody molecule, modifying the antibody into a shape capable of combining with that particular antigen. In this "instructive" model, the lymphocyte only gains its ability to produce a specific antibody after its initial contact with antigen. In 1955, Niels Jerne, a Danish immunologist, proposed a radically different mechanism. Jerne suggested that the body produces small amounts of randomly structured antibodies in the absence of any antigen. As a group, these antibodies could combine with any type of antigen to which a person might someday be exposed. According to Jerne's model when a person is exposed to an antigen,

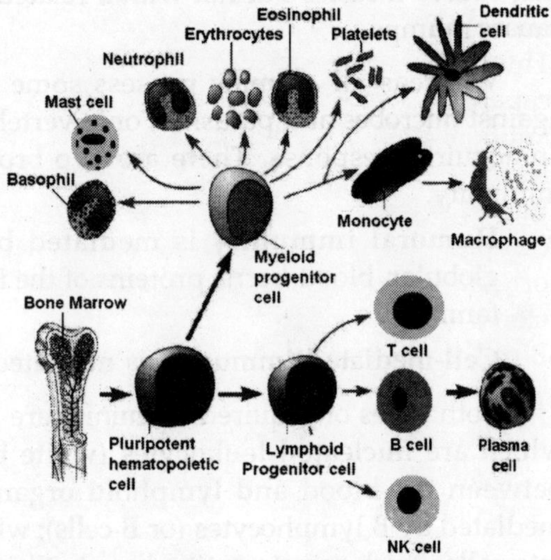

Figure 1.3 Pathways of differentiation of pluripotent haematopoietic stem cells in the bone marrow. *A haematopoietic stem cell can give rise to two different progenitor cells: A myeloid progenitor cell that differentiates into erythrocytes, basophils, neutrophils, macrophages and dendritic cells, or lymphoid progenitor cells that differentiate into any of the various types of lymphocytes (NK cells, T cells, or B cells). T cell precursors migrate to the thymus where they undergo differentiation into T cells. In contrast, B cells undergo differentiation in the bone marrow.*

the antigen combines with a specific antibody and leads to the subsequent production of many molecules of that particular antibody. Thus in Jerne's model, the antigen selects those pre-existing antibodies capable of binding to it. In 1957, the concept of antigen selection of antibodies was expanded into a comprehensive model of antibody formation by the Australian immunologist F. MacFarlane Burnet. Burnet's clonal selection theory has been widely accepted and remains compatible with all the experimental evidence accumulated since its original proposal. An overview of the steps that occur during the clonal selection of B cells is shown in **Figure 1.4**. The main features of B-cell clonal selection are as follows:

1. **Each B cell becomes committed to produce one species of antibody**. B cells arise from a population of undifferentiated stem cells. (**Figure 1.3**) It becomes committed to producing only one species of antibody molecule as the result of random DNA

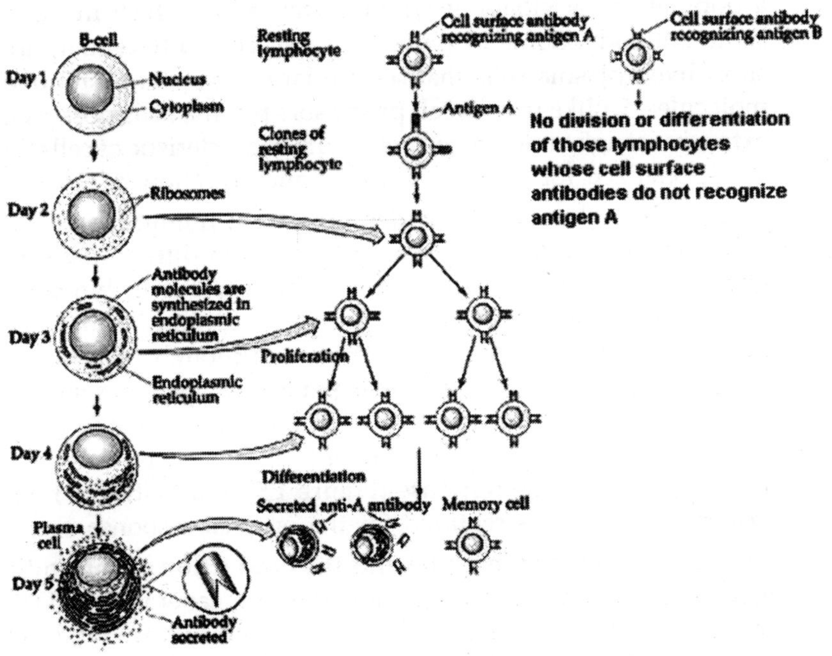

Figure 1.4 ***The clonal selection of B cell by a thymus-independent antigen.*** *B lymphocyte recoginizes antigen (A) through its specific cell surface receptor and proliferates to form many plasma cells producing anti-A antibodies. Some cells differentiate into memory cells which retain the specificity for the antigen. B cell bearing receptor for antigen B fails to proliferate in response to antigen A.*

rearrangements. Since thousands of different rearrangements are possible, B cells develop that are capable of producing a large variety of antibody molecules. Each B cell therefore, bears a unique surface immunoglobulin molecule, which is termed "B cell receptor" (BCR). However, it must be remembered that an individual B cell can only produce one type of antibody.

2. **Antibody production follows selection of B cells by antigen.** Activation of a B cell by antigen requires the involvement of T cells, except in case of polysaccharides antigens, which can activate B cells independent of T cells (thymus independent antigens). On exposure to *Haemophilus influenzae* type B, an encapsulated bacterium that causes fatal meningitis, the capsular polysaccharide of the organism binds to a small fraction of the body's B cells. These B cells bear membrane-bound antibodies with specific combining site for the antigen. This receptor ligand interaction activates the B cell, causing it to proliferate and form a population (or clone) of lymphocytes, all of which make the same antibody. Some of these activated cells differentiate into short-lived plasma cells that secrete large amounts of antibody molecules. Unlike their B cell precursors plasma cells possess an extensive rough endoplasmic reticulum characteristic of cells that are specialized for protein synthesis and secretion.

3. **Immunologic memory provides long-term immunity**. Not all B-lymphocytes that are activated by antigen differentiate into antibody-secreting plasma cells. Some remain in lymphoid tissues as memory B cells that can respond rapidly at a later date on exposure to the same antigen. Although plasma cells are short lived, memory B cells may persist for a person's lifetime. On subsequent contact with the same antigen, some of the memory B cells rapidly proliferate into plasma cells, generating a secondary immune response (**Figure 1.5**) in a matter of hours rather than the days required for the original response.

4. **Immunological tolerance prevents the production of antibodies against self**. Genes encoding antibodies are generated by DNA rearrangement, a process in which DNA segments are randomly combined. As a result of the extremely random nature of rearrangement, B cells that recognize body's own tissues are also generated. These have the capability to produce widespread organ destruction and subsequent disease. It is obviously in the best interest of the body to prevent the production of such Ig

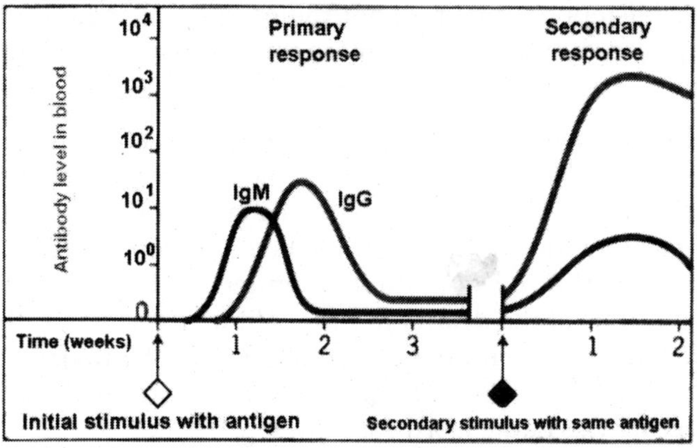

Figure 1.5 Primary and secondary antibody responses. *A primary response is elicited by initial exposure to an antigen leading to the production of soluble IgM antibody molecules followed by production of soluble IgG antibody molecules. Appearance of blood-borne antibodies follows a delay of several days. When the antigen is reintroduced at a later time, a secondary response is initiated. In contrast to the primary response, the secondary response begins with production of IgG (as well as IgM) molecules and leads to a much higher antibody level in the blood and occurs with almost no delay.*

molecules, which are called autoantibodies. During the development of the immune system, cells capable of recognizing auto antigens are eliminated by apoptosis (a process called clonal deletion) or inactivated and become unable to respond to antigen (a process called clonal anergy). As a result, the body develops immunologic tolerance toward itself. However, all cells capable of producing autoantibodies are not eliminated and hence in some individuals autoimmune diseases, such as lupus erythematosus and rheumatoid arthritis develop.

Several principles of the clonal selection theory and primary and secondary antibody responses can be illustrated by briefly considering the subject of vaccination.

Vaccination

Edward Jenner practiced medicine in the English countryside at a time when smallpox was one of the most prevalent and dreaded diseases. Over the years, he noticed that maids who tended cows were typically spared the ravages of the disease. Jenner concluded that milkmaids were somehow "immune" to smallpox because they were infected at an early age with cowpox, a harmless disease they contracted from cows. Cowpox produce blisters that resemble the pus-filled blisters of smallpox but the cowpox blisters are localized and disappear, causing nothing more serious than a scar at the site of infection.

In 1796, Jenner performed one of the most famous (and risky) medical experiments of all time. First, he infected an eight-year-old boy with cowpox and gave time to the boy to recover. Six weeks later, Jenner intentionally infected the boy with smallpox by injecting pus from a smallpox lesion directly under the boy's skin. The boy showed no signs of the deadly disease. Within a few years, thousands of people had become immune to smallpox by intentionally infecting themselves with cowpox. This procedure was termed vaccination, after "vacca," the Latin word for cow.

Jenner's smallpox vaccination produced immunity by stimulating T cells. Most of the vaccines currently in use are B-cell vaccines, such as that employed to fight tetanus. Tetanus results from infection by the anaerobic soil bacterium *Clostridium tetani*, which can enter the body through a puncture wound. As they grow, the bacteria produce a powerful neurotoxin that blocks transmission across inhibitory synapses on motor neurons, leading to sustained muscle contraction and asphyxiation. In early infancy, infants are immunized against tetanus by inoculating with a modified and harmless version of the tetanus toxin (called a toxoid). The tetanus toxoid binds to the surface of B cells whose membrane-bound antibody molecules have a complementary binding site. These B cells proliferate to form clones of cells that produce antibodies capable of binding to the actual tetanus toxin. This initial response soon wanes, leaving some memory cells that respond rapidly upon re-exposure to *C. tetani* infection at a later date. The immunity to the tetanus toxoid does not always last a lifetime, which is why a booster dose of the vaccine is recommended.

T lymphocytes: Activation and mechanism of action

Like B cells, a process of clonal selection also operates for T cells. T cells possess a cell-surface protein, called T-cell receptor, which allows them to interact with a specific antigen. Like the B-cell receptors, the T-cell receptors also exist as a large population of molecules that have differently shaped combining sites. Just as each B-cell produces only one species of antibody, each T-cell has only a single species of T-cell receptor. It is estimated that adult humans possess approximately 10^{12} T cells that, collectively, exhibit all the antigen receptors required to deal with external agents.

Unlike B cells, which are activated by soluble, intact antigens, T cells are activated by fragments of antigens that are displayed on the surfaces of other cells, called antigen-presenting cells (APCs). If a liver or kidney is infected with a virus, infected cells would display portions of the viral proteins on its surface, enabling the infected cell to bind to a T-cell with the appropriate T-cell receptor. As a result of this presentation, the immune system becomes alerted to the entry of a specific pathogen.

Professional antigen presenting cells (APC) include dendritic cells and macrophages, which play a key role in initiating an immune response. Dendritic cell (DCs), which are often described as the "sentinels" of the immune system, have earned this title because they "stand guard" in the body's peripheral tissues where pathogens are likely to enter (such as the skin and airways). They also accumulate at sites of infection and inflammation. As immature cells in the body's tissues, DCs are engaged primarily in the uptake of materials by endocytosis and phagocytosis. Once an antigen is taken into a dendritic cell, it has to be processed before it can be presented to another cell. DCs that have processed antigen migrate to nearby lymph nodes where they differentiate into mature APC's. In the lymph node, DCs come into contact with a large pool of resident T cells, including the rare T-cells which have TCRs that can bind specifically with the processed foreign antigen. Binding of the antigen to the TCRs activates these T-cells, which proliferate and form a clone of cells having the same T-cell receptor. It is estimated that a single activated T-cell can divide as many as 15 times in less than a week, generating a large population of T cells capable of interacting with the foreign antigen. The massive proliferation of specific T lymphocytes in response to an infection is often reflected in the enlargement of local lymph nodes. Once the foreign antigen has been cleared, the vast majority of the expanded T-cell population die by programmed cell death or apoptosis. A small fraction of the cells is retained as memory T cells, which are capable of responding rapidly in the event of future contact with the same pathogen.

T cells mediate their function in close contact with other cells, which include APCs, B cells, other T cells or target cells located anywhere in the body. In addition, cells secrete highly active chemical messengers, called cytokines, that at very low concentrations mediate many of these interactions. Cytokines are small proteins produced by a wide variety of cells and include interferons (IFNs), interleukins

(ILs), and tumour necrosis factor (TNF). Cytokines bind to specific receptors on the surface of a responding cell, generating an internal signal that alters the activity of the cell. In responding to a cytokine, a cell may prepare to divide, undergo differentiation or secrete its own cytokines. One family of small cytokines, called chemokines, acts as chemo attractants by stimulating the migration of leukocytes. Different types of lymphocytes and phagocytes possess receptors for different chemokines, so that their migration patterns can be separately controlled. A list of some of the best-studied cytokines is given in **Table 1.1**.

Table 1.1 Selected cytokines

Cytokine	Source	Major functions
IL-1	Diverse	Induces inflammation, stimulates T_H- cell proliferation.
IL-2	T_H cells	Stimulates T – cell and B –cell proliferation.
IL-4	T cells	Induces IgM to IgG class switching in B cells, suppresses inflammatory cytokine action.
IL-5	T_H-cells	Stimulates B-cell differentiation.
IL-10	T cells	Inhibits macrophage function, suppresses inflammatory macrophages cytokines action.
IFN-γ	T_H-cells CTLs	Induces MHC expression in APCs, activates NK cells.
TNF-α	Diverse	Induces inflammation, activates NO production in macrophages.
GM-CSF	T_H-cells CTLs	Stimulates growth and proliferation of granulocytes and macrophages.

Cells of the immune system

Mononuclear Phagocytes: These cells have a common lineage from a hematopoietic precursor in bone marrow and their primary function is phagocytosis. These cells enter the circulation as incompletely differentiated monocytes, which are 10-15 um in size and have an eccentric bean-shaped nucleus, fine granular cytoplasm containing lysosomes, phagocytic vacuoles and cytoskeletal filaments. These migrate into tissues and mature into macrophages that take different morphological forms at different sites e.g. microglial cells of central nervous system, Kupffer cells lining the vascular sinusoids in liver, alveolar macrophages in lung. These cells are important mediators of innate immunity since they can phagocytose almost any substance. However, they also play a crucial role in adaptive immunity since they can display foreign antigens on their surface in a form that can be recognized by antigen specific T lymphocytes thus giving them the name "antigen presenting cell" (APC). The macrophages that perform this function carry specialized plasma membrane glycoprotein's called major histocompatibility complex (MHC) class II molecules. The processing of foreign antigens by macrophages and other APCs involves four stages. First, antigens are taken into the cell by a relatively non-specific form of endocytosis that brings molecules dissolved in the extracellular fluid into the cell. Next, the internalised endocytic vesicles fuse with liposome's, whose hydrolytic enzymes degrade the foreign antigens into small fragments in the case of ingested protein antigens; the process yields peptide fragments about 10-20 amino acid long. Third, the resulting fragments become bound to MHC class II molecules, which possess an antigen-biding cleft that is specifically designed to bind antigen fragments. And finally, the foreign antigen- MHC II complex is transported to the cell surface for "presentation" to other cells of the immune system. The importance of the preceding pathway has been demonstrated by treating cells with inhibitors that block endocytosis or the lysomal degradation of antigens; under such conditions, an immune response is inhibited. Endogenous antigens that originate within the cells of the host organism can also trigger immune responses. For example, cells that have become infected by a virus usually manufacture foreign proteins encoded by viral genes. Fragments derived from the viral proteins become bound to a different type of MHC molecule, called MHC class I molecule. Unlike class II MHC molecules, which are expressed only on cells of the immune system, class I molecules are expressed by almost all nucleated cells. Peptide antigens destined to bind to class I MHC molecules are produced by protein cleavage in the cytosol and are then transported into the ER lumen, where the fragments bind to class I molecules associated with the ER membrane. The resulting antigen MHC complex is then transported to the cell surface, where its presence can trigger an immune response against the infected cell.

Dendritic Cells: These are morphologically distinct cells with membranous or spine-like projections and are of two types. The interdigitating dendritic cells are located in the interstitium of most organs, are abundant in the T-cell rich areas of lymph nodes and spleen and are scattered throughout the epidermis of skin where they are called Langerhans cells. These cells arise from marrow precursors and are closely related in lineage to macrophages. They are very efficient antigen presenting cells. The follicular dendritic cells (FDC) are located in the germinal centres of lymph nodes, spleen and mucosa associated lymphoid tissues. These are not of bone-marrow origin, are unrelated to the interdigitating dendritic cells and display "trapped" antigens on their cell surface for recognition by B lymphocytes.

Granulocytes: Granulocytes are cells, which have a granular cytoplasm, are stimulated by T-cell derived cytokines and mediate inflammation. They are of three types, neutrophils, eosinophils and basophils, based on the staining property of the granules. Neutrophils respond to chemotactic stimuli and are the first cells to reach the site of infection where they phagocytose and destroy foreign particles. Eosinophils are found mostly at sites of allergic responses, which are IgE mediated. Basophils are effectors of immediate hypersensitivity responses.

Lymphocytes: These are the most important cells of the immune system, which mediate the adaptive responses. They consist of T and B-cell populations and the T cells are further subdivided into cytotoxic and helper T lymphocytes based on cell surface receptors and biological functions.

Cytotoxic T lymphocytes (CTLs or T_C cells) are T cells which recognise and kill target cells bearing antigenic fragments alongwith class I MHC. Unlike the NK cells which forms part of the innate immune response, CTLs kill target cells in an antigen specific manner as part of the adaptive response. CTLs kill target cells by inducing them to undergo apoptosis. The CTL releases ***perforins*** and ***granzymes*** into the space between the cells. *Perforins* are proteins that assemble within the membrane of the target cell to form transmembrane channels. *Granzymes* are proteolytic enzyme that enter the *perforin* channels and activate caspases, which are the proteolytic enzyme that initiate the apoptotic response. By killing infected cells, CTLs eliminate viruses, bacteria, yeast, protozoa, and parasites after they have found their way into host cells and are no longer accessible to circulating antibodies. CTLs possess a surface protein called CD8 and are referred to as CD8+ cells.

Helper T lymphocytes (T_H cells) are distinguished from CTLs by the presence of the protein CD4 at their surface rather than CD8. T_H cells are activated by professional APCs. Once activated T_H cells regulate subsequent immune responses by recognizing and activating other lymphocytes that are specific for the same antigen. Nearly all B cells require the help of T_H cells before they can mature and differentiate into antibody-secreting plasma cells. B cells are activated by direct interaction with a T_H cell, as shown in **Figure 1.6**.

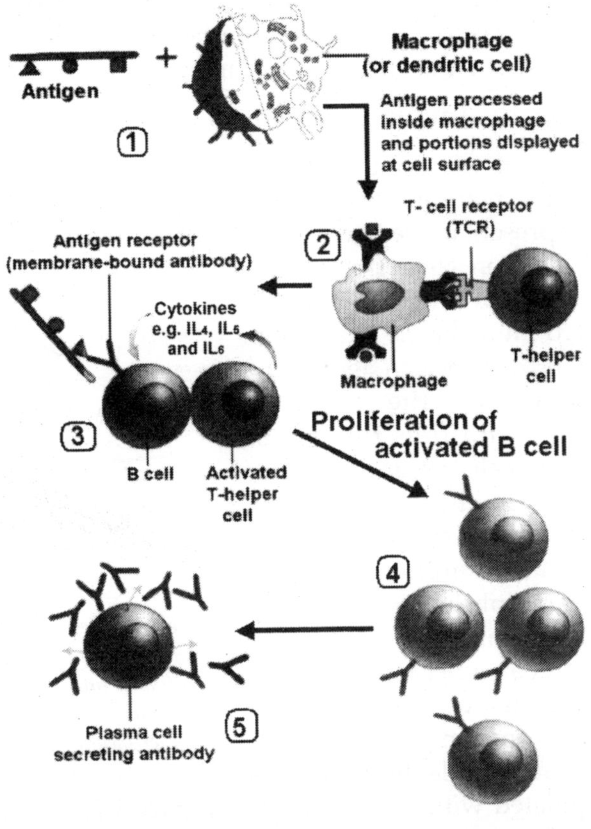

Figure 1.6 Schematic drawing showing the role of T_H cells in antibody formation. *(1) Macrophage interacts with complex antigen and is taken into the macrophage and cleaved into fragments, which are displayed at the cell surface. (2) Macrophage interacts with a T_H cell whose TCR is bound to one of the displayed antigen fragment and activates T-cell (3) Activated T_H cell interacts with a B-cell whose antigen receptor is bound to an intact, soluble antigen. B-cell activation is stimulated by cytokines (e.g., IL-4, IL-5, and IL-6) released by the T_H cell into the space separating it from the adjacent B-cell. Interaction with the T_H cell activates the B-cell, causing it to proliferate (4) Progeny of the activated B-cell differentiate into plasma cells that synthesize antibodies that can bind the antigen.*

Thus, the formation of antibodies requires the activation of both T cells and B cells capable of interacting specifically with the same antigen. The importance of T_H cells becomes apparent when one considers the devastating effects caused by human immunodeficiency virus (HIV), the virus that causes AIDS. T_H cells are the primary targets of HIV, which enters a host CD_4+ cell by binding to its surface protein. Most HIV-infected people remain symptom free for as long as their T_H cell count remains above 500 cells per mm (the normal count is over 1,000 cells per mm). Once the count drops below approximately 200 cells per mm, infected person develops full-blown AIDS and becomes susceptible to viral infections.

The Organs of the immune system

The cells of the immune system are present as organized structures in the bone marrow, thymus, lymph nodes and spleen, besides being present in the circulation. The cells move freely between organs and their presence throughout the body allows the immune response to be activated immediately at any site, on stimulation by a foreign agent. The bone marrow is the site of haematopoiesis and thus the site of origin of all the cells except follicular dendritic cells. The thymus lies in the anterior mediastinum and is the site of maturation of T lymphocytes. It is divided into multiple lobules by fibrous septae and each lobule consists of an outer cortex and an inner medulla; non-lymphoid epithelial cells are scattered throughout the organ. The cortex is densely populated while the medulla is sparsely populated with T lymphocytes (**Figure 1.7**). The thymus is the site of the "clonal selection" of T lymphocytes, the process, which results in the production of mature T lymphocytes that recognize only non-self antigen.

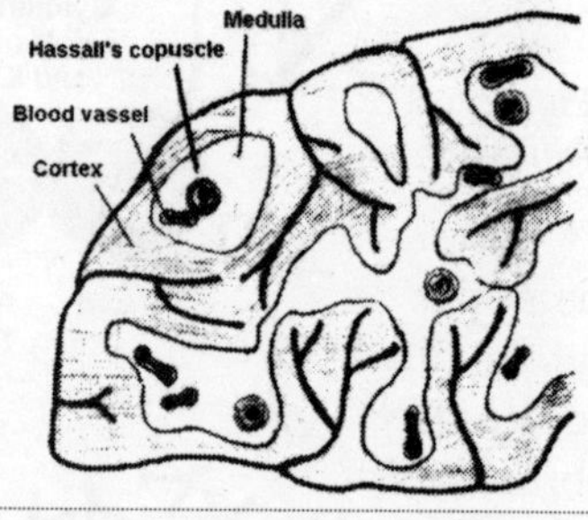

Figure 1.7 Schematic diagram of the thymus. *Illustrating a part of a lobe divided into multiple lobules. Each lobule consists of outer cortex and inner medulla.*

Lymph nodes are small nodular aggregates of lymphoid tissue situated throughout the body. Most organs have a rich supply of lymphatics, which drain into these nodes.

These organs have a dense fibrous capsule, an outer cortex

containing lymphoid follicles (aggregation of lymphocytes) and an inner medulla. The follicles are rich in B-lymphocytes while T lymphocytes are located in the paracortical areas and the mantle zone around follicles (**Figure 1.8**).

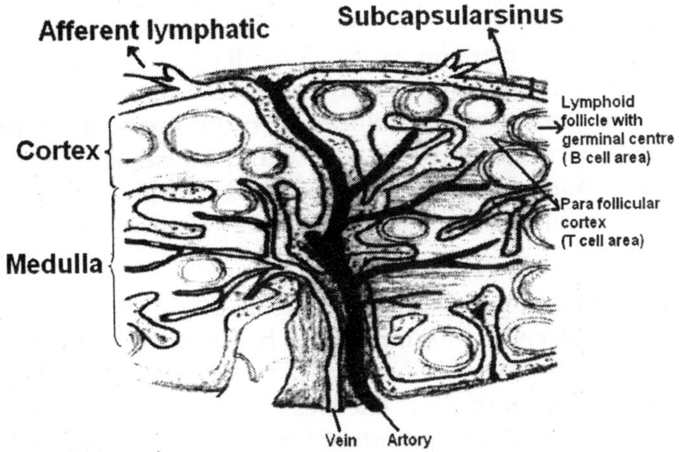

***Figure 1.8 Schematic diagram of a lymph node** showing the cortex with lymphoid follicles and dense lymphocytes and the medulla with lymphatic cords and vessels.*

Spleen is located in the upper left quadrant of the abdomen and is approximately 150g in weight in adults. The splenic artery enters at the hilum to divide into progressively smaller branches. The cellular collection around these branches is the red pulp of the spleen and is mostly made of T lymphocytes. Interspersed between these is the white pulp. B lymphocytes are predominantly located in the follicles and germinal centres (**Figure. 1.9**).

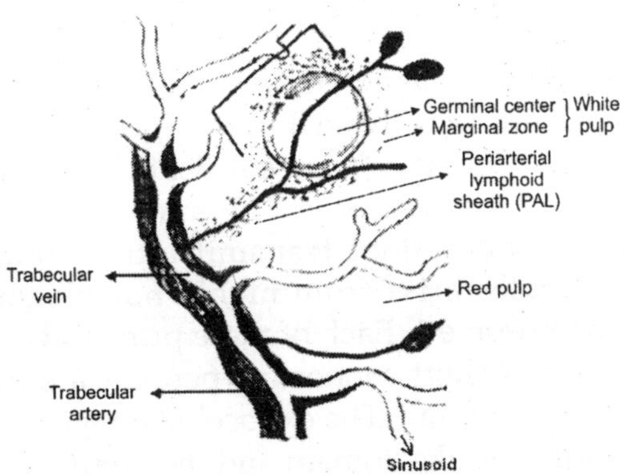

*Figure 1.9 **Schematic diagram of the spleen.** The white pulp is made of dense lymphoid tissues in periarteriolar sheaths and follicles; it is intermingled with the red pulp which is composed of vascular sinusoids and scattered cells.*

Cellular co-operation in the immune response

CD4+ T lymphocytes are activated when antigen is presented to them for recognition on the surface of the specialized antigen presenting cells (APCs). The Langherhans cells in the skin migrate from the skin via the efferent lymphatic into the paracortical areas of the draining lymph nodes where they interdigitate with many T-cell. This migration provides an efficient mechanism for carrying antigen from skin to the CD4+ T cells located in the lymph nodes. The APCs are rich in class II major histocompatibility complex molecules (MHC), which are important for presenting antigens to the CD4 cells. The dendritic cells, which are found in the secondary follicles of B cell areas of lymph nodes and spleen, present antigen to B cells. These specialized cells internalize antigen by phagocytosis or endocytosis and then express a part of that antigen together with the MHC molecule on their membrane.

The major histocompatibility complex (MHC) is located on short arm of chromosome 6 in humans and has 2 sets of genes which are responsible for the two biochemically distinct sets of proteins, class I and class II MHC antigens. The class II molecules are cell surface heterodimers, which are important for cellular co-operation because the T-cell receptor only recognizes antigen if it is presented on the surface of APCs within an MHC class II molecule.

The antigen presenting cells utilize their enzymatic machinery to break down complex antigens into short peptides. These peptides are taken up by the MHC class II groove and transported to the cell surface. The MHC Class II peptide complex is recognized by the appropriate TCR on T cells.

The engagement of TCR by the MHC-peptide complex is not sufficient to activate T cells. The CD3 molecule that consists of five chains ($\gamma,\delta,\varepsilon,\zeta,\eta$) is non-covalently associated with the α β heterodimer of the T-cell receptor. The intracellular region of the TCR chains is short and cannot transmit intracellular signals. Chains of CD3 molecules arise from highly homologous set of genes located on chromosome 1. Each of these proteins has an N-terminal extracellular region, a short connecting peptide, a transmembrane segment and a cytoplasmic tail. The extracellular region of these 3 molecules contains a single Ig-like domain and therefore the three peptides are members of the Ig super family. The cytoplasmic domains of the CD3, ε and ζ proteins range from 44 to 81 amino acids in length and are therefore efficient in transducing signals to the cell interior.

The ζ and η chains are encoded by alternatively spliced RNA transcripts of the same gene resulting in two proteins that have identical extracellular and transmembrane segments but differ in their cytoplasmic tails, which are of 113 and 155 amino acids in length. Tight co-regulation of the expression of TCR peptides and the CD3 ε and ʒ peptides are required for the surface expression of all these structures. When the TCR is successfully engaged by the MHC / peptide, the associated CD3, ε and ʒ /h chains transduce the signals to the cytoplasm of T cells which lead to functional activation. However, the CD3 and ʒ /h molecules do not themselves associate with the MHC at any point.

An additional molecule on T cells, which binds to a non-polymorphic site on the MHC molecule, is the CD4/CD8 peptide. These glycoprotein which are expressed on mutually exclusive subsets of T cells, have structural limitations in their association with APC. Class I MHC binds only to CD8 + T cells while class II MHC binds only to CD4+ T cells. This gives a functional bifurcation to T-cell recognition in that the T-helper functions of CD4+ cells is strictly dependent on recognition of the CD4 molecule by the class II molecule on APCs -a phenomenon termed "restriction". Similarly, the cytotoxic function of CD8+ T cells is class I "restricted". The distinct pathway which delivers the peptides to MHC class I molecules processes most intracellular antigens, which include viruses. These are recognised by CD8+ cell, which are cytotoxic in nature and will lyse the presenting cells since this is the only way intracellular antigens can be eliminated. Since viruses can infect any cell in the body, it is advantageous that class I molecules are present on all nucleated cells. Extracellular antigens are processed and presented on the class II MHC molecule to be recognised by CD4+ cells. This stimulates a helper response with antibody production, which is useful to neutralise or lyse extracellular organisms by a variety of mechanisms.

Besides the CD4/CD8 association with MHC molecule, additional interactions between T cells and accessory cells take place. This makes cell-to-cell interaction stronger and at least one of them, CD28-B7, is essential as a second signal for T cell activation. The binding of TCR with antigen without B7-CD28 interaction results in cellular signals, which permanently switch off the cellular machinery, a situation which results in anergy of the cell. Adequate expression of B7 -CD28 is therefore essential for T-cell activation.

Once adequate surface engagement takes place and intracellular signals are initiated, cascades of activation pathways are triggered, which finally results in up regulation and secretion of various cytokines. These play a role in further expansion of the effector responses such as maturation and differentiation of B cells to produce antibodies. Recruitment of naïve cells as well as multiplication of committed cells contributes to the expansion of effector responses. During this phase, some of the progeny of multiplying cells become long-lived memory cells. These cells help in early activation of a secondary response.

The whole cascade of events is exquisitely orchestrated and initial antigenic challenge only produces an immune response that is adequate to control this challenge. This implies that the expansion of the effector response is not uncontrolled but regulated to meet the needs of the challenge. The regulation occurs at multiple stages and is less well understood. It involves regulatory T cells, soluble suppressor factors, idiotype networks and other possible mechanisms, which are beyond the scope of this chapter. An inefficient regulation results in an uncontrolled immune response that could be harmful to the host.

A highly evolved immune system that is activated by millions of different antigens raises the important question: How do organisms prevent harming themselves by directing immune responses against their own self antigens and normal cells? In other words, how does the immune system distinguish "self" from "non-self"? Ray Owen, who discovered that non-identical twin cattle occasionally exchange blood cells during embryonic development and retain these "foreign" cells into adulthood, provided an early clue in 1940s. If they had been injected directly into adult cattle, such foreign cells would have been attacked and destroyed by the immune system. This and other similar observations led Macfarlane Burnet to propose that antigens encountered by the immune system during embryonic development are recognized as self-antigens, inducing a permanent state of immunological tolerance. Peter Medawar, injected embryos of one mouse strain with tissues derived from another strain. When the animals reached adulthood, they were unable to reject skin grafts from mice whose tissues they had encountered during embryonic life, although their immune responses were otherwise normal.

What causes animals to become "tolerant" of antigens encountered during embryonic development? Burnet's clonal deletion theory has postulated that self-recognizing lymphocytes are eliminated during embryonic development when the immature lymphocytes encounter these components.

During early embryonic development a heterogeneous population of immature T cells expressing many million different TCRs arise in the thymus. These immature cells, which serve as precursors for both helper T cells and killer T cells, carry both CD4 and CD8 on their surface. As the first step in T-cell maturation, immature lymphocytes that bear TCRs recognizing self MHC molecules are selected for further maturation, a process termed "positive selection". Subsequently cells bearing TCRs that bind to self -antigen fragments associated with either class I or class II MHC molecules are eliminated by a phenomenon termed "negative selection". By deleting such cells before they can mature into helper or killer T cells, the organism protects itself from future attack by its own immune system. Depending on whether the TCR binds to a class I or class II MHC molecule, the selected cell either loses CD4 and develops into a CD8+ cytotoxic T-cell, or it loses CD8 and develops into a CD4+ helper T cell.

Although clonal deletion is now an established mechanism, it represents only one of the several mechanisms that appear to be involved in immune regulation.

Since clonal deletion is not perfect, cells that escape are held in check in the periphery by the mechanism of "peripheral tolerance". Immune escape occurs when organisms are exposed to antigens under conditions that do not trigger an immune response. Lymphocytes shut themselves down instead of becoming activated if they encounter a foreign antigen that has not been properly presented by an antigen-presenting cell. Thus self-reactive cells stay inactive, except in an aberrant situation when the antigen gets presented in the appropriate context.

Failure of immunological tolerance can trigger autoimmune diseases

Immune tolerance is the mechanism that deletes or suppresses lymphocytes that are potentially harmful because their antigens receptors are targeted against an organism's own molecules. When these mechanisms fail, immune responses directed against normal healthy tissues could trigger the onset of severe "autoimmune" diseases. Such diseases in humans include multiple sclerosis, in which the myelin surrounding nerve axons is attacked, rheumatoid arthritis, in which the joints are targeted, type I juvenile diabetes, in which insulin producing cells are destroyed, myasthenia gravis, in which the acetylcholine receptor is targeted and systemic lupus erythematosis, in which multiple organs including blood vessels, skin, and kidneys are attacked.

Important insights into the nature of autoimmunity have emerged from studies involving experimental autoimmune encephalomyelitis (EAE). Injecting animals with a basic protein derived from the myelin sheath that surrounds nerve axons can induce EAE The injected protein triggers an immune response directed against myelin, leading to a widespread paralysis that is often fatal. Lymphocytes taken from affected animals proliferate ***in vitro*** when stimulated with the offending antigen. The proliferating lymphocytes were identified as helper T cells whose antigen receptors were specific for myelin basic protein. These T cells could passively transfer disease phenotype into normal rats. When T-cell clones directed against myelin basic protein was inactivated by X-irradiation, and then injected back into a normal animal, the animals failed to develop EAE. Further injections of myelin basic protein also failed to induce the disease. The resistance was because injected T cells acted as an antigen, triggering an immune response directed against the antigen binding site of the TCRs carried by these T cells. Animals acquired lasting protection against any lymphocyte whose antigen receptor binds to myelin basic protein.

CHAPTER 2

Introduction of Basic Genetics

Human Genetics

Strategies for determining the inheritance of human traits differ markedly from those used to study the genetics of peas, fruit flies, roundworms, corn or mice. Humans have small families and long generation times. Mendelian proportions cannot be expected among the progeny of a single human marriage (mating). There is no possible way - taking into account ethical moral and practical considerations - that human mating can be controlled on the basis of the genotype of the parents.

Yet genetic analysis of human traits can be studied effectively by two approaches. Data from large populations can be collected and mathematical methods applied to deduce whether a particular trait is inherited. Although adequate, this approach requires the examination of many individuals, and is hence arduous, expensive and time consuming. Alternately, the pattern of inheritance of a particular trait can be followed within families, which should preferably be large, extended multi-generation families. The examination of such families is often easier to conduct than population studies. Both approaches are widely used for investigating the molecular genetics of human diseases.

Calculation of Mendelian frequencies

Monogenic traits are also known as Mendelian traits, so named after Gregor Mendel, a Nineteenth Century Austrian monk who deduced several important genetic principles from his well-designed experiments with garden peas. Mendel studied seven traits in the pea, each of which is determined by a single gene. These traits included attributes such as height (tall plants vs. short plants) and seed shape (rounded vs. wrinkled). The variation in each of these traits is caused by different alleles at individual loci.

Two central principles emerged from Mendel's work. The first is the principle of segregation, which states that sexually reproducing organisms possess genes that occur in pairs and that only one member of this pair is transmitted to the offspring (i.e. it segregates). The prevalent thinking during Mendel's time was that hereditary factors from two parents are "blended" in the offspring. In contrast, the principle of segregation stated that genes remain intact and distinct. An allele for "rounded" seed shape can be transmitted to an offspring who can, in turn, transmit the same allele to its own offspring. If, instead of remaining distinct, genes were somehow blended in offspring, it would be impossible to trace genetic inheritance from one generation to the next. Thus, the principle of segregation was a key development in modern genetics.

Mendel's principle of independent assortment is his second great contribution to genetics. This principle states that genes at different loci are transmitted independently. Consider the two loci mentioned previously. One locus can have either the "rounded" or wrinkled" allele and the other can have either the "tall" or the "short" allele. In a reproductive event, a parent will transmit one allele from each locus to its offspring. The principle of independent assortment dictates that the allele transmitted at one locus ("rounded" or "wrinkled") has no effect on which allele is transmitted at the other locus ("tall" or "short").

Mendel's experiments also showed that the effects of one allele might mask those of another. He performed crosses (mating) between pea plants homozygous for the "tall" gene (i.e., they had two identical copies of an allele labelled as H) and plants homozygous for the "short" gene (having two copies of an allele labelled as h). This cross, which can produce heterozygous (Hh) offspring, is illustrated in the Punnett square shown in **Figure 2.1**. Mendel found that the offspring of these crosses, even though they were heterozygotes, were all tall. This reflects the fact that the H allele is dominant, while the h allele is recessive (it is conventional to label the dominant allele in upper case and the recessive allele in lower case). The term "recessive" comes from a Latin root meaning "to hide". This is an appropriate description for the behaviour of recessive alleles since in heterozygotes, the consequences of a recessive allele are hidden. Whereas a dominant allele exerts its effect both in the homozygote (HH) and heterozygote (Hh), the presence of the recessive allele is detected only when it occurs in homozygous form (hh). Thus, short pea plants could be created only by crossing parent plants that each carry at least one h allele. An example would be a heterozygote X heterozygote cross, shown in **Figure 2.2**.

Parent

Parent

	h	h
H	Hh	Hh
H	Hh	Hh

Figure 2.1 Punnett square illustrates a cross between HH and hh homozygote parents.

Parent

Parent

	H	h
H	HH	Hh
h	Hh	hh

Figure 2.2 Punnett square illustrates a cross between two Hh heterozygotes.

The principle of segregation describes the behaviour of chromosomes during meiosis. The genes on chromosomes segregate during meiosis and they are transmitted as distinct entities from one generation to the next. When Mendel performed his critical experiments, he had no direct knowledge of chromosomes, meiosis, or genes (indeed, the latter term was not coined until 1909, long after Mendel's death). Although his work was published in 1865 and cited occasionally, its fundamental significance was unrecognised for several decades. Mendel's research, which was eventually replicated by other researchers at the turn of the twentieth century, forms the foundation of modern genetics.

The patterns of inheritance of a trait within families are visualized by constructing family trees (pedigrees). The term pedigree is an English idiom for the French phrase *pied de grue,* which means "foot of a crane" and was a whimsical way for the French to describe the lines of a genealogical chart. A pedigree used for genetic studies is depicted using a set of symbols describing both individual relationships and the history of a trait within a single or extended family (**Figure 2.3**). Pedigrees are essential experimental data for human geneticists.

A reliable set of clinical features that accurately categorize a human phenotype (condition) is essential for determining whether the condition is inherited or not. If the description of a condition is imprecise, then different genetic and non-genetic disorders may be grouped together, there by confusing analysis. Since the nineteenth century, clinicians have sought to create exact, inclusive descriptions of each human disease. A consistent, definitive set of signs and symptoms that accompanies a condition is commonly called a syndrome. Some syndromes are easy to diagnose, whereas others require a more detailed medical assessment.

For some studies, many different pedigrees must be assembled. In others, a single, large multi-generation pedigree may provide extensive genetic information. With unambiguous information, a human geneticist selects from only four different familial patterns to determine the mode of inheritance of a trait that resides at a single gene locus. Included in this category of single- gene (monogenic) traits are autosomal dominant, autosomal recessive, X-linked dominant, and X- linked recessive conditions.

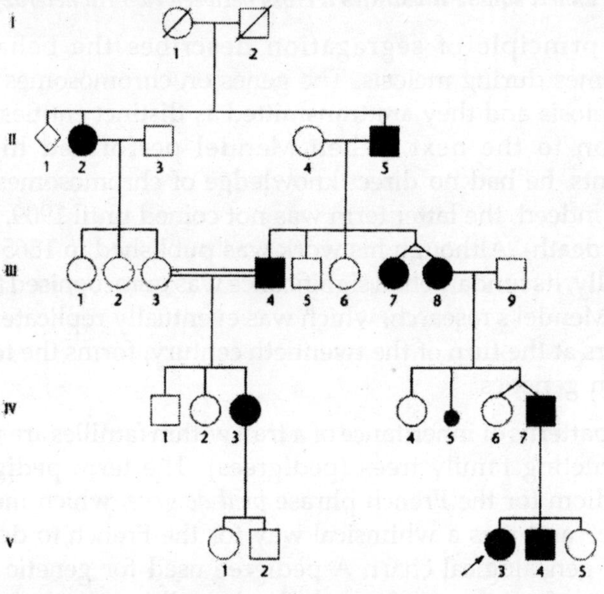

Figure 2.3 Interpreting standard pedigree notations. *Circles represent females and squares represent males. Completely solid (filled in) symbols mark affected individuals. Detailed description of various symbols is given in Figure 2.11.*

Autosomal Dominant inheritance

There are approximately 200 human genetic conditions known to be caused by dominant genes with loci disbursed among all the autosomes. These autosomal dominant (ad, AD) disorders can affect any organ system and occur with different frequencies (**Table 2.1**).

A typical pedigree of an autosomal dominant condition has distinguishing features, which include the presence of affected individuals in successive generations, equal numbers of affected males and affected females, each affected individual usually having at least one affected parent, two affected parents having unaffected offspring, and unaffected individuals of the pedigree who marry unaffected individuals rarely having affected offspring (**Figure 2.4**).

Table 2.1 A List of some human autosomal dominant disorders

Disorder	Prevalence	Description
Huntington's disease	1/10,000	Late onset; degeneration in the cerebral cortex and basal ganglia; involuntary movement (chorea); dementia
Neurofibromatosis type 1	1/5,000	Multiple neurofibromas on the nerves of the head, neck and body; pigmented (café-au-lait) spots
Tuberous sclerosis	1/5,800	Multisystem; growths (haematomas) within the brain, eyes, skin, kidneys, heart, lungs, and skeleton
Myotonic dystrophy	1/8,000	Multisystem; prolonged muscle contraction (myotonia); variable muscle weakening and wasting (atrophy); cataracts; defective impulse conduction by the heart; inadequate gonadal function (hypogonadism)
Polycystic kidney disease	1/1,000	Genetically heterogeneous; variable age of onset; kidney cysts; decreased kidney concentrating ability; enlarged kidneys; hypertension
Retinitis pigmentosa	1 /4,000	Genetically heterogeneous; progressive loss of night vision and visual acuity

Disorder	Prevalence	Description
Marfan's syndrome	1/10,000	Multisystem; long fingers and toes (arachnodactyly); skeletal deformities; loose joints; ocular lens dislocation; impaired vision; cardiovascular disorders; lateral curvature of the spine (scoliosis); rupture of the aorta
Waardendurg syndrome	1 / 1,00,000	White forelock; premature graying; different colured eyes; deafness
Hypercholesterolemia	1 / 500	High serum cholesterol levels; early – onset coronary artery disease
Osteogenesis imperfecta	1 / 10,000	Genetically heterogeneous; clinically heterogeneous; bone deformity; brittle bones; deafness; blue sclerae

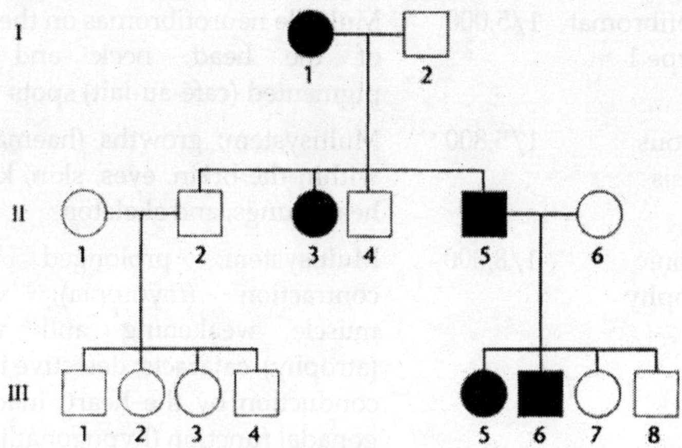

Figure 2.4 A sample pedigree with features of an autosomal dominant

Autosomal recessive inheritance

Approximately 900 different autosomal recessive (AR) disorders affect a range of organ systems and the genes responsible for them are scattered among all the autosomes. Two defective alleles at the same locus cause these disorders. Some human autosomal recessive disorders are shown in **Table 2.2.**

Table 2.2 A list of some human autosomal recessive disorders

Disorder	Prevalence	Description
Cystic fibrosis	1/ 2,500	Multisystem; defective chloride transport in epithelial tissue; blockage of ductules and small airways; severe lung disease; pancreatic insufficiency; chronic inflammation of the nasal sinuses (sinusitis); infertility
Lamellar ichthyosis	1 / 25,000	Disorder of the skin; disfigurement; large scales; variable redness
Wilson's disease	1/ 40,000	Chronic liver disease; progressive neurological impairment; accumulation of copper in the liver, brain, and other tissues
Gaucher's disease type 1	1/ 50,000	Increased blood cell breakdown in the spleen (hypersplenism); enlargement of the liver and spleen (hepatosplenomegaly); fragile bones
Friedreich's ataxia	1/50,000	Onset at puberty; inability to co-ordinate muscles for voluntary movement (ataxia); inability to speak (dysarthria); muscle atropy
Childhood spinal muscular atrophy	1/ 10,000	Genetically heterogeneous; variable; degeneration of the anterior horns of the spinal cord; weakness and wasting of the muscles; often fatal by 20 years of age or earlier
Phenylketonuria	1/ 10,000	Deficiency of liver enzyme phenylalanine hydroxylase; brain damage; mental retardation; excess phenylpyruvate in the urine; accumulation of phenylalanine in the blood
□- Thalassemia	1/ 20,000	Severe depletion of red blood cells (anemia); enlarged spleen (splenomegaly); bone deformities
Galactokinase deficiency	1/40,000	Inability to utilize galactose; cataracts; mild mental retardation
□-1 – Antitrypsin deficiency	1/ 3,500	Breathlessness (emphysema); cirrhosis of the liver

The distinctive features of a pedigree with an autosomal recessive disorder include unaffected parents who have affected offspring, equal numbers of affected males and females, all offsprings affected when both parents are affected, and in rare disorders, affected children who are the offspring of marriages between first cousins or other closely related family members (**Figure 2.5**).

The interpretation of the mode of inheritance from a single pedigree is not always reliable. If a defective recessive allele is frequent, then homozygous recessive individuals have a high probability of marrying heterozygotes. In these cases, an autosomal recessive condition may mimic an autosomal dominant pattern. Usually, the nature of the mode of inheritance becomes apparent after pedigrees from many different families with the same condition are assembled and examined.

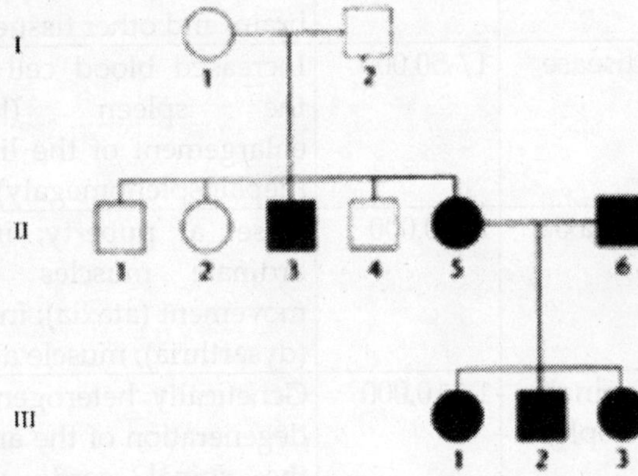

Figure 2.5 A sample pedigree with features of an autosomal recessive disorder.

X- linked inheritance

In fertile human males, half the sperm carry a Y chromosome and the other half an X chromosome. All unfertilised eggs of a female receive a single X chromosome. Consequently, within a population, half of the next generation will be XX (females) and the other half will be XY (males) (**Figure 2.6**). The segregation of the sex chromosomes during meiosis is an effective way to maintain a sex ratio close to 1:1 from one generation to the next. With this sex determination system, all sons inherit a Y chromosome from their fathers. In humans, an important gene for determining testis formation (SRY, sex determining region of the Y chromosome) is located at chromosome Yp11.3.

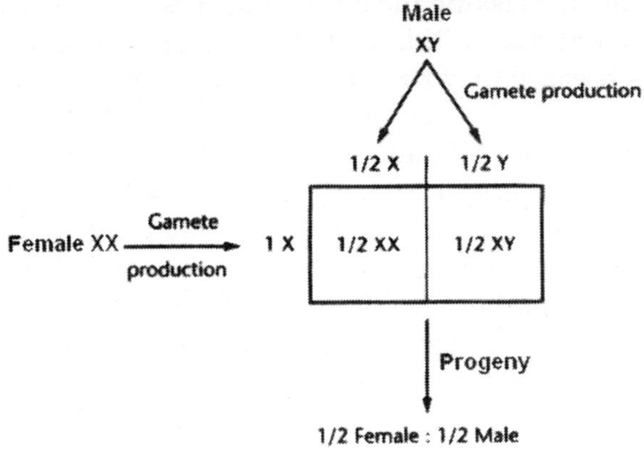

Figure 2.6 Distribution of sex chromosomes and determination of the sex ratio in humans.

Except for two segments (at least), which are designated pseudoautosomal regions (**Figure 2.7**), the sex chromosomes do not share much chromosomal material. The pseudoautosomal regions are locations where synapsis and crossing over occur between the X and Y chromosomes. Synapsis of the X and Y chromosome ensures the correct segregation of the sex chromosomes during meiosis. Both synapsis and crossing over are more common between the Xp-Yp (1-1) than the Xq – Yq (2-2) pseudoautosomal regions. Approximately 23 genes have been localized to the non-pseudoatuosomal portion of the Y-chromosome. These Y – linked genes have been called holandric genes. The genetic make-up of the X chromosome is well known. More than 285 X-linked genes have been identified.

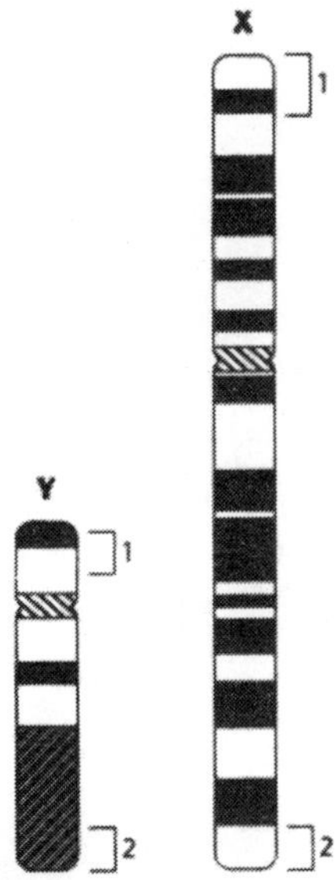

Figure 2.7 Human sex chromosomes. *The numbered brackets mark the approximate locations of two pseudo-autosomal regions. The Xp – Yp pseudoautosomal regions (1 –1) synapse more frequently than the Xq – Yq pseudoautosomal regions (2 – 2).*

A single X-linked recessive allele is sufficient to produce a phenotype in human males, because there is little homologous genetic information shared between the X and Y chromosomes. Consequently, in these cases, there is virtually no chance of a normal dominant allele masking the effect of a recessive allele. The dominant or recessive effect of an X-linked gene is determined by the phenotype in females because they have two X chromosomes. In pedigrees with an X-linked dominant condition, each generation usually has an affected individual. All daughters of affected males are affected, both sons and daughters of an affected heterozygous female may be affected, and generally, twice as many females as males are affected (**Figure 2.8**). In pedigrees with X-linked recessive disorders, all sons of an affected mother are affected, affected fathers never transmit the trait to their sons, unaffected parents may have affected offspring, and generally, there are more affected males than females (**Figure 2.9**).

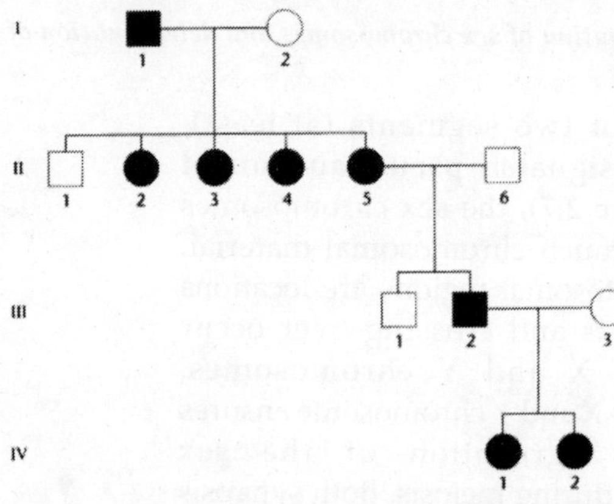

Figure 2.8 ***A sample pedigree with features of an X-linked dominant inherited disorder***

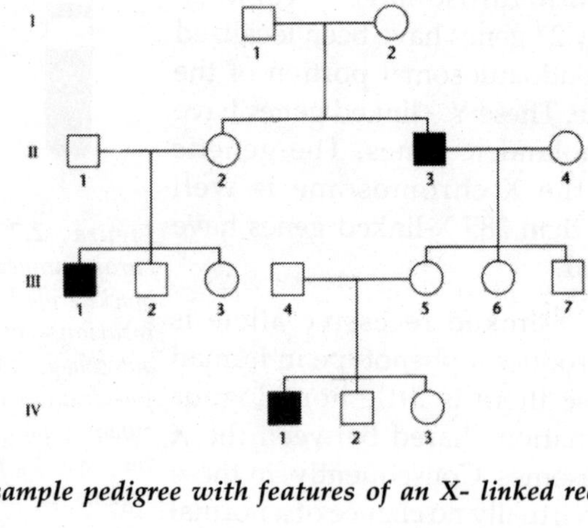

Figure 2.9 ***A sample pedigree with features of an X- linked recessive inherited disorder.***

X inactivation

In early 1960s Mary Lyon hypothesized that one of X chromosome in each somatic cell of the female is inactivated. This provides dosage compensation that is an equalization of X linked products in males and females. The **Lyon hypothesis** states that X inactivation occurs early in female embryonic development and that the X chromosome contributed by the father is inactivated in some cells, whereas in other cells the X chromosome contributed by the mother is inactivated. This inactivation procedure is random, so that maternally and paternally derived X chromosomes will each be inactivated in about half of the embryo cells. Once an X chromosome is inactivated in a cell, it will remain inactive in all descendants of the cell. X inactivation is, therefore, said to be randomly determined, but fixed. As a result of X inactivation, all normal females have two distinct populations of cells with active paternally or maternally derived X chromosome. (**Figure 2.10**). Hence, females are mosaics for the X chromosome. Male having only one copy of the X chromosome, are not mosaics but are hemizygous for the X chromosome.

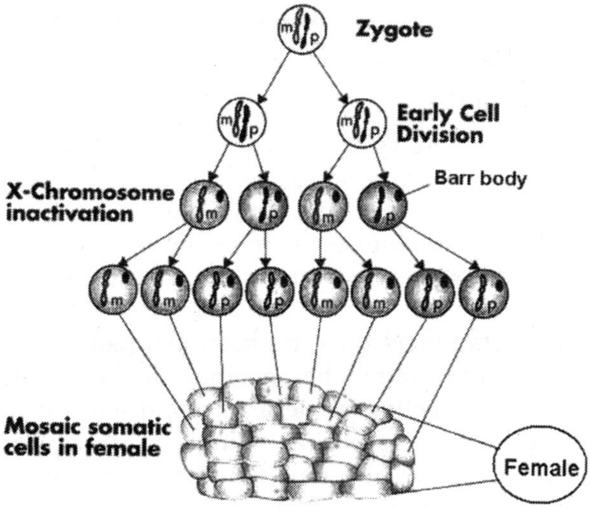

Figure 2.10 The X inactivation process. *The maternal (m) and paternal (p) X chromosome are both active in the zygote and in early embryonic cells. X inactivation then takes place resulting in cells having either an active paternal X or an active maternal X. Females are thus X chromosome mosaics, as shown in the tissue sample at the bottom of the figure.*

Genes on the Y-chromosomes

Very few genes have been localized to the human Y chromosome. This failure to find Y-linked genes is somewhat surprising, since a mutation in one of these genes should have an immediate phenotypic effect on the man who carries it. Furthermore such a mutation should be passed on to all his sons but to none of his daughters. A Y – linked gene should therefore be the easiest kind of gene to identify in conventional pedigree analysis. To date, however, only a few Y-linked genes have been discovered. One is responsible for the synthesis of male-specific substance called H-Y antigen, which is found on the cell surface. Another is involved in the production of a factor that is critical for the differentiation of the testes and the subsequent acquisition of male sexual characteristics. The Y-chromosome has fewer genes than any other human chromosome.

Genetic Heterogeneity

A genetic disorder, which may appear to be a single entity, is frequently found to be genetically heterogeneous on closer analysis. That is, it may include phenotypes that are similar but are actually determined by different genotypes. Genetic heterogeneity may be the result of different mutations at the same locus (allelic heterogeneity), mutations at different loci (locus heterogeneity), or both.

Locus Heterogeneity

Genes having more than one discernible effect on the body are said to be pleiotropic. A good example of a gene with pleiotropic effect is Marfan syndrome. First described by Antoine Marfan, a French pediatrician, this autosomal dominant disorder affects the eye, skeleton and cardiovascular system. All the observed features of Marfan syndrome are due to unusually stretchable connective tissue. The gene for Marfan syndrome has been mapped chromosome 10q 15q. It has been demonstrated that mutations in the encoding fibrillin, a component of connective tissue are responsible for the multiple defects seen in this disorder just as a single gene may have multiple effects, a single disease phenotype may be caused by mutations at different loci in different families. The causation of the same disease phenotype by mutations at distend loci is termed locus heterogeneity. A good example is that of adult pholysystic kidney disease (AKPD), an autosomal dominant disorder in which a progressive accumulation of renal cysts are seen patients may also develop liver cysts, hypertension, cerebral aneurysms, and cardiac valvular defects.

Ocuring in about 1/1,000 caucasians, this disorder is responsible for 8% to 10% of end stage renal disease in North America. APKD can be caused by mutation in genes on either chromosome 16 (PKD1) or chromosome 4 (PKD2). Both these genes encode membrane spanning glycoproteins that interact with one another and may be involved in cellular signaling (when this signaling goes away, it is thought that cellular growth regulation is compromised, resulting in cyst formation). In one family the disease may be caused by a PKD1 mutation, whereas in another family it may be caused by a PKD2 mutation and these two causes may be clinically indistinguishable.

Ostegenesis imperfecta provides a second example of locus heterogeneity. The procollagen triple helix is encoded by two genes on chromosome 7 and 17. A mutation in either of these genes can after the normal structure of the triple helix, resulting in osteogenesis imperfecta.

Allelic heterogeneity

Allelic heterogeneity is an important cause of clinical variation. Many loci possess more than one mutant allele; in fact, at a given locus there may be several or many mutations. Sometimes, these different mutations result in clinically indistinguishable or closely similar disorders. In other cases, different mutant alleles at the same locus result in very different clinical presentations. For example, some mutations in the RET gene, which encodes a receptor tyrosine kinase, can cause dominantly inherited failure of development of colonic ganglia leading to defective colonic motility and severe constipation (Hirschsprung disease), whereas other mutations in the same gene result in dominantly inherited cancer of the thyroid and adrenal glands (multiple endocrine neoplasia type IIa and IIb). A third group of mutations in RET causes both Hirschsprung disease and multiple endocrine neoplasia in the same individuals.

Unless they have consanguineous parents, most people with autosomal recessive disorders are more likely to have compound rather than truly homozygous genotypes. However, they are usually referred to as homozygotes if they have a pair of abnormal alleles at a locus. Because different allelic combinations may have somewhat different clinical consequences, clinicians must be aware of allelic heterogeneity as one possible explanation for variability among patients considered to have the same disease.

To establish the pattern of transmission, a usual first step is to obtain information about the family history of the patient and to summarize the details in the form of a pedigree; a graphical representation of a family tree, using standard symbols is shown in **Figure 2.11**.

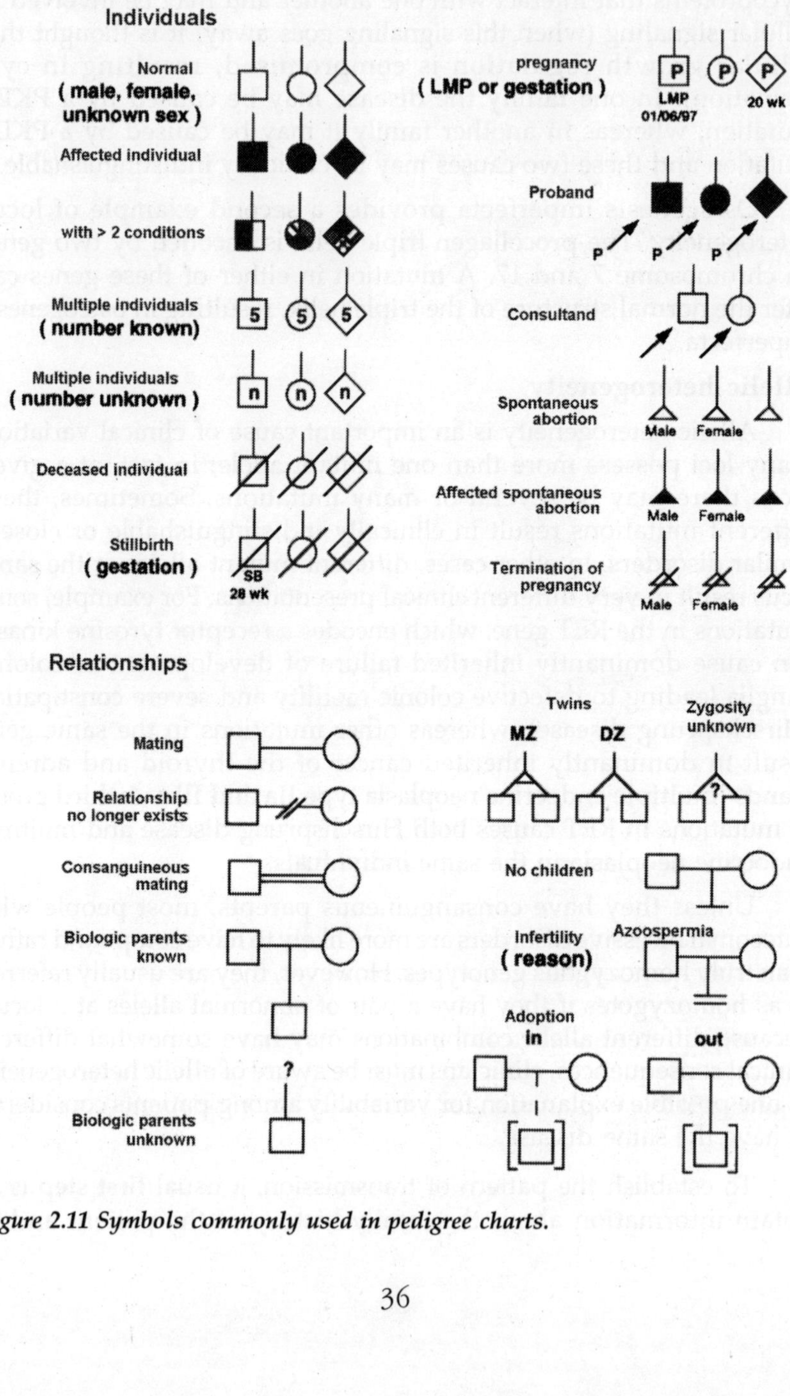

Figure 2.11 Symbols commonly used in pedigree charts.

Atypical patterns of inheritance

As a general rule, the pattern of inheritance and segregation ratios of single-gene disorders is in accordance with the principles of Mendelian inheritance. Throughout the 20th century, since Garrod and Bateson first applied Mendel's laws to inborn errors of metabolism, few exceptions to Mendelian inheritance have been observed. However, close examination of certain unusual disorders and the analysis of mutations in molecular detail has shown that exceptions to Mendelian inheritance do occur in single-gene disorders and must be considered in genetic medicine. The divergence from a typical pattern of inheritance could be because of genetic heterogeneity, locus heterogeneity, allelic heterogeneity or linkage.

The study of a disorder often begins when an individual with an unusual set of symptoms is referred to a clinician, who assesses the array of abnormalities and compares them with those of known syndromes. If there is a close match with an existing syndrome, then it is assumed that the condition is identified. If the signs of the condition are novel, further study may be undertaken to determine whether the condition is inherited or the result of an environmental factor.

The immediate family members of the affected individual (proposita, propositus, proband) found to have a new or interesting condition are examined, and a pedigree is constructed. The pedigree is enlarged, if possible, by studying family archives, interviewing additional family members, and having others fill in questionnaires. The pattern of inheritance may be revealed at this stage, but a single pedigree, unless it is quite large (more than 50 members) and has many affected individuals, usually does not provide sufficient information to deduce the mode of inheritance. If the initial pedigree does not produce a definitive genetic interpretation, the clinician may seek other families with the same condition. The mode of inheritance should become evident after a number of different pedigrees have been assembled.

However, analyses of genetic disorders are not always straightforward and in some instances, different pedigrees with the same set of clinical symptoms may show completely different modes of inheritance as was previously discussed for retinitis pigmentosa. Genetic heterogeneity is the term used to describe different, non-allelic loci that produce the same phenotype. When faced with genetic heterogeneity, which appears to be more the rule than the exception, clinicians look for subtle physiological and physical differences that may distinguish the effect of one locus from that of others. In this way, the overall phenotype is often divided into various subtypes.

It is seldom apparent why different genes cause the same syndrome. However, if one considers, developmental processes or the proper functioning of organ systems to be similar to metabolic pathways, there are a number of sequential steps that are required for a biological end result. Each step is mediated by the product of a different gene and loss of any one of these steps prevents the end result from being realized. Hence many non-allelic mutant genes that are involved in the same pathway could produce the same phenotype.

Other factors may confound pedigree analysis. In some instances, although an individual has a certain genotype, the expected phenotype is not produced. This phenomenon is called incomplete penetrance. More specifically, penetrance is a quantitative measure of the frequency of occurrence of the phenotype when a particular genotype or allele is present. With most genetic disorders, each time an individual has a specific genotype, the phenotype is evident. This represents complete (100%) penetrance. For other conditions, the penetrance of a genotype is incomplete (less than 100%). For example, if 87 out of 112 people with the same genotype show the typical phenotype while the remainder do not, then the penetrance for this disorder is 78% [(87 /112) x 100] and non-penetrance is 22% {[(112 - 87) / 112] x 100}.

Another confounding aspect of some genetic disorders is that the phenotype may not express at birth. The age of onset of a disorder can range over decades, and with some genetic diseases, the condition does not appear until the fourth, fifth, or sixth decade after birth. Disparity in the age of onset of a disorder can cause members of a pedigree to be wrongly classified at the time the pedigree is assembled.

The same allele in different individuals may give rise to diverse phenotypes. For example, the autosomal dominant disorder, osteogenesis imperfecta (OI), is a condition that causes brittle bones, loose joints, fragile skin, deafness, soft teeth, and a blue tint to the outer layer (sclera) of the eyeball. Within a single family, the phenotype can vary considerably, with some affected family member having only blue sclerae, whereas another is almost totally incapacitated because of brittle bones. This type of phenotypic variation is called variable expressivity, variability or clinical heterogeneity. Whereas, incomplete penetrance may result in a particular genotype having no apparent effect on some individuals, clinical heterogeneity results in some aspect of the phenotype being observed in all individuals with the same genotype, although with variable degrees of severity.

There are no simple explanations for incomplete penetrance, age-dependent penetrance, or clinical heterogeneity. Incomplete penetrance may be the result of the activity of a gene product of a non-allelic gene locus nullifying the effect of the disease-causing allele. Age - dependent penetrance may result over a period of time from cumulative tissue damage from an altered gene product. In these cases, the definitive clinical features that mark a disorder may not be evident until a certain level or threshold of tissue damage or cell death is reached or surpassed. Clinical heterogeneity can occur if non-allelic gene products modify the consequences of a disease - causing allele. Regardless of the basis of these confounding factors, they add complexity to the interpretation of inheritance of disease using pedigree analysis.

Calculations of probability

Risk assessment is an important part of medical genetics. For example, the physician or genetic counsellor has to frequently inform couples of the risk of having a child with a genetic disorder. In order to understand how such risks are estimated, we need to know some basic concepts of probability. Probability is defined as the proportion of times that a specific outcome occurs in a series of events. Thus, we may speak of the probability of obtaining a 4 when a die is tossed, or the probability that a couple will produce a son versus a daughter. Since probabilities are proportions, they lie between 0 and 1, both inclusive.

During meiosis, one member of chromosome pair is transmitted to sperm or egg. The probability that one member of the pair is transmitted is ½ , and the probability that the other member of the pair is transmitted is also ½ (note that the probabilities of all possible events must add to 1 for any given experiment). Since this is directly analogous to coin tossing, in which the probabilities of obtaining heads or tails are each ½ we will use coin tossing as our illustrative example.

When a coin is tossed repeatedly, the outcome of each toss has no effect on subsequent outcomes. The events are thus said to be independent. Event if we have obtained ten heads in a row, the probability of obtaining heads or tails on the next toss remains ½. Similarly, the probability that a parent will transmit one of the two alleles at a locus is independent from one reproductive event to the next.

The independence principle allows us to deduce two fundamental concepts of probability, the multiplication rule and the addition rule. The multiplication rule states that if two trials are independent, then the probability of obtaining a given outcome in both trials is the product of the probabilities of each outcome. For example, we may wish to know the probability that an individual will obtain heads on both tosses of a coin. Because the tosses are independent events, this probability is given by the product of the probabilities of obtaining heads in each individual toss: ½ x ½ = ¼. Similarly, the probability of obtaining two tails in a row is ½ x ½ = ¼ .

The multiplication rule can be extended for any number of trials. Suppose that a couple wants to know the probability that all three of their planned children will be girls.Since the probability of producing a girl is ½, and since reproductive events are independent of one another, the probability of producing three girls is ½ x ½ x ½ = 1/8. However, if couple already has two girls then the probability of having a third girl, is only ½ since the previous two events are no longer probabilities. Because of independence, events that have already occurred have no effect on the outcome of the third event.

The addition rule states that if we want to know the probability of either one outcome or another, we can simply add the respective probabilities together. For example, the probability of getting two heads in a row (½ x ½ or ¼) or the probability of getting two tails in a row (1/4) is given by adding the two probabilities together: ¼ + ¼ = ½. If a couple plan to have three children, and have a strong aversion to having all three children of the same sex, they can be reassured some what by knowing that the probability of producing three girls (1/8) or three boys (1/8) is only ¼ (1/8 + 1/8). The probability that they will have some combination of boys and girls is thus ¾ , since the sum of the probabilities of all possible outcomes must add to 1.

Gene and genotype frequencies

The prevalence of many genetic diseases can vary considerably from one population to another. For example, cystic fibrosis, a severe respiratory disorder is quite common among Caucasians, affecting approximately 1/2500 births. It is rare in Asian populations, affecting only 1/90000 births. The sickle cell disease is common among African -Americans, affecting approximately 1/600 births. Yet it is almost never seen among individuals of north European descent. The concepts of genotype frequency and gene frequency help us to measure and understand population variation in disease genes.

Imagine that 200 individuals in a population have been tested for the MN blood group. This blood group, which is encoded by a locus on chromosome two, has two major alleles, labelled M and N and the effects of both alleles can be observed in the heterozygotes. M and N are thus said to be co- dominant and the heterozygotes can be distinguished from both homozygotes. Any individual in the population can have one of the three possible genotypes, homozygous for M (genotype MM), heterozygous (MN), or homozygous for N (NN). After typing all the individuals in our sample the following distribution of genotypes was found, MM, 64; MN, 120; NN, 16. The genotype frequency is obtained by dividing each genotype count by the total number of subjects. Thus, frequency of MM is 64/200 (=0.32), frequency of MN is 120/200 (=0.60) and the frequency of NN is 16/ 200 (=0.08). The sum of these frequencies must of course equal 1.

The gene frequency for each allele, M and N, can be obtained by the process of gene counting. Each MM homozygote has two M alleles, while each heterozygotes has one M allele. Similarly, NN homozygotes have two N alleles and heterozygotes have one N allele. Hence, in the above example, there are 248 M genes [(64 x 2) + 120] and 152 N genes [(16 x 2) + 120].

There are 400 genes at the MN locus (i.e., two alleles per subject) and gene frequency of M, 248/400 = 0.62 and of N, 152/400 is 0.38. The sum of the two frequencies must equal 1.

Linkage

Thomas Hunt Morgan used the fruit fly *Drosophila melanogaster* in 1910 - 11 to study crosses between females heterozygous at two loci, the genes purple and vestigial, affecting eye colour and wing shape, with males who were homozygous for each mutant gene (**Figure 2.12**). The mutant alleles are written pr and vg respectively and the "wild type" alleles are written pr+ and vg+.

The reason is that the two genes under consideration are located on the same chromosome (**Figure 2.12**) and the parental combinations of alleles remain together unless separated by a recombination (**Figure 2.13**).

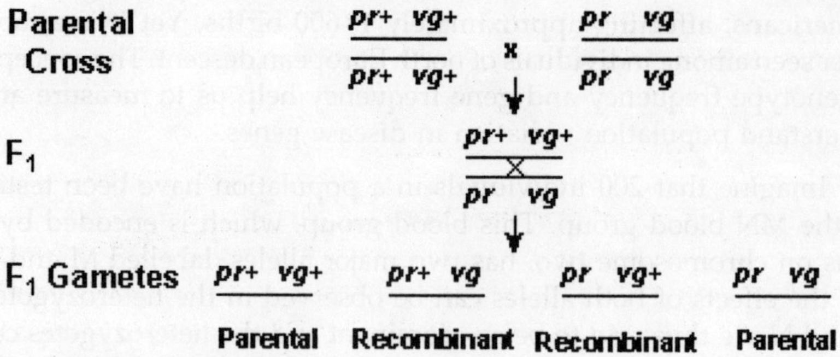

Figure 2.12 Morgan's First Cross

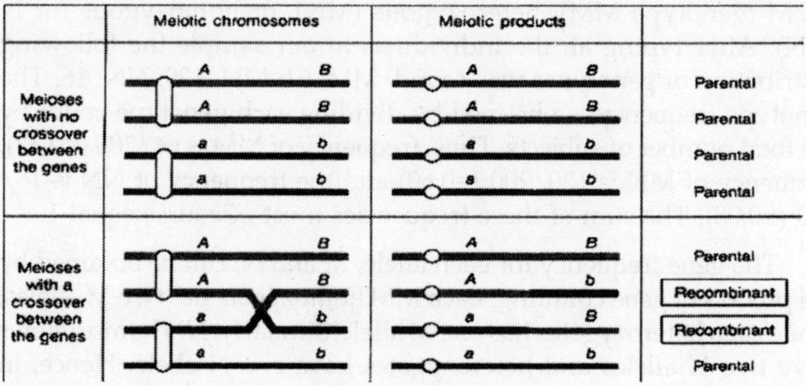

Figure 2.13 Parental combinations and recombination

The closer the two genes are, the more likely they are to be inherited in the parental combination as shown in the **Figure 2.14.** This can be quantified by defining the "Recombination Fraction". q = (number of recombinant progeny) / (total number of progeny) . q must lie in the range from 0 to 0.5, where value of 0 implies that the two genes are so close to each other that they never recombine and a value of 0.5 implies that the genes are unlinked. The maximum value is 0.5 and not 1 because crossing over takes place after DNA replication and involves only one chromatid per chromosome of the pair.

The linkage map

Since genes are organized along a linear stretch of the DNA molecule, measuring the frequencies of recombination between genes could create a map. The recombination frequency is the measure of the genetic distance between loci and a map unit corresponds to 1% recombination. This is named as centi Morgan in honour of Thomas Hunt Morgan. The purple and vestigial genes described above are 10.7cM apart.

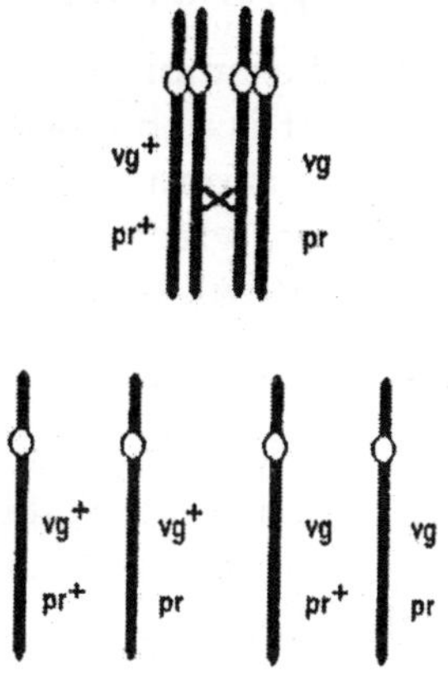

Figure 2.14 Crossover: When there is a crossover, half the resulting gametes are still non-recombinant. So even if a crossover occurs in every

Lethal alleles

Some dominant mutations are, in fact, semi-dominant and the phenotype of the homozygotes is more extreme than the phenotype of the heterozygote. The normal allele of the gene T (Denforth's short tail) in mice is expressed in the embryo; T/+ mice develop a short tail but T/T homozygotes die at early embryonic stage. Laboratory stocks are maintained by crossing the heterozygote as shown below

T/+ x T/+

|

V

T/T		T / +		+/+	
1	:	2	:	1	ratio at conception
0	:	2	:	1	ratio at birth

Incomplete dominance

In incomplete dominance (some time called partial dominance), a heterozygous genotype creates an intermediate phenotype. In this case, both the dominant and recessive genes are expressed, creating a blended or combined phenotype. A cross of two intermediate phenotypes can result in the reappearance of either the parent phenotype or the blended phenotype.

The classical example of this is the colour of carnations. Let us suppose R is the gene for red pigment R′ is the gene for no pigment. Offspring with the genotype RR will have prominent pigmentation, R′R′ offspring will have no pigment and appear white, RR′ and R′R offsprings will have a little bit pigment and appear pink.

An example of incomplete dominance in human is of mordan, a trait that is exhibited when eye colour alleles from maternal and paternal chromosomes are blended. This usually occurs when one parent has green eye colour and other parent has brown eye colour and child will have dark blue eyes.

Co-dominance

If two or more alleles can be distinguished in the phenotype in the presence of the other they are said to be co-dominant. An example is seen in the ABO blood group where A and B alleles are co-dominant (**Table 2.3**)

Table 2.3: ABO genotypes and phenotypes

Genotype	Phenotype	Red cell antigens	Serum antibodies
AA	A	A	Anti- B
AO	A	A	Anti-B
BB	B	B	Anti-A
BO	B	B	Anti-A
AB	AB	A and B	Neither
OO	O	Neither	Anti-A and anti - B

The ABO genes codes for a glycosyl-transferase that modifies the H antigen on the surface of red blood cells. The O allele has a frame shift mutation in the gene and thus produces a truncated and inactive product which cannot synthesize H. 'A' phenotype individuals have natural antibodies to B antigen in their serum and vice versa. O phenotype individuals have antibodies directed against both A and B. AB individuals have no antibodies against either A or B antigens.

Epistasis

This occurs where the action of one gene masks the effects of another making it impossible to predict the genotype at the second gene. Such a situation may occur when two genes produce enzymes, which act in the same biochemical pathway.

If product of gene 1 is not present because the individual is homozygous for a mutation, it will not be possible to tell what the genotype is at gene 2.

Pleiotropy

Pleiotropy refers to mutations selected for their effect on one specific character which often affect other characters of the organism. This might mean that there are related physiological pathways contributing to a similar phenotype in several tissues. For example, the white colour mutation in Drosophila results in the lack of pigmentation not only in compound eyes but also in ocelli (simple eyes), sheaths of tissues surrounding the male gonad and the malpighian tubules (the fly's kidneys). In all these tissues pigment formation requires the uptake of pigment precursors by the cells. The white allele causes a defect in this uptake, there by locking pigment formation in all these tissues.

Often pleitropy involves multiple events that are not obviously physiologically related. For example, the dominant Drosophila mutation Dicha etc. causes the wings to be held out laterally but also removes certain hair on the back of the fly; furthermore, the mutation is lethal when homozygous. This example shows a real limitation in the way dominant and recessive mutations are named. The reality is that the same mutation can be dominant and recessive depending upon which aspect of its pleiotropic phenotype is under consideration.

Variable expressivity

The degree to which a disease may manifest itself can be very variable, as in tuberous sclerosis in which some individuals scarcely have any symptom, whereas others are severely affected. Very mild symptoms may in fact get overlooked and if wrongly classified as non-affected, could have profound implications for genetic counselling.

Incomplete Penetrance

This is an extreme case of a low level expressivity when some individuals who ought to show symptoms because of their genotype, do not. In such cases, even the most careful clinical examination reveal no symptoms and individual may be misclassified normal until the defective gene gets transmitted and the offspring manifests the disease.

Anticipation

In some genetic diseases, symptoms get progressively worse in each successive generation as shown in **Table 2.4**. The autosomal dominant condition, myotonic dystrophy, is characterized by myotonia, muscular dystrophy, cataract, hypogonadism, frontal

Table 2.4: Myotonic dystrophy

Generation	Number of CTG repeats in	Phenotype
1	5	Normal
2	19 - 30	"pre-mutant"
3	50 - 100	Mildly affected
4	200 or more	Severely affected

balding and ECG changes and is caused by the expansion of a trinucleotide repeat in the 3' untranslated region of a gene on chromosome 19. The severity of the disease roughly correlates with the number of copies of the trinucleotide repeat unit.

The "premutant" individuals have a small expansion of the number of trinucleotide repeats which is insufficient to cause any clinical effect in itself but it allows much greater expansions to occur during the mitotic divisions before gametogenesis. Mildly affected individuals can again have gametes in which a second round of expansion might have occurred. An expansion of the CTG triplet repeat in the 3′ untranslated end of the myotonic dystrophy gene occurring predominantly in maternal meiosis appears to be the explanation for the severe neonatal form of myotonic dystrophy which usually only occurs when the gene is transmitted by the mother. A similar expansion in the 5′ end of the Huntington's disease gene is paternal meiosis appears to account for the increased risk of Juvenile Huntington's disease when the gene is transmitted by the father.

Mosaicism

An individual, or a particular tissue of the body, can consist of more than one cell type or line, through an error occurring during mitosis at any stage after conception, a situation that is known as mosaicism. Mosaicism of either somatic or germ cells can account for some instances of unusual patterns of inheritance.

Somatic mosaicism

Somatic mosaicism is suspected when the features of a single gene disorder are less severe in an individual than usual or confined to a particular part of the body in a segmented distribution. If the mutation arises early during development and is present in all or some of the germ cell lines, it can be transmitted to the next generation with full expression.

Germ line Mosaicism

If a new mutation occurs in one germ cell precursor out of the many non-mutant precursors, its descendent germ cells, being diluted by many non-mutant germ cells present, will not produce mutant offspring in the expected Mendelian numbers.

Phenocopies

This refers to effects of environmental factors, which may mimic a genetic trait. For instance a heat shock delivered to Drosophila pupae may cause a variety of defects, which mimic those caused by mutations in genes affecting wing or leg development. In humans, the drug thalidomide taken during pregnancy caused phenocopies of the rare genetic disease; phocomelia and children were born with severe limb defects.

Exceptions to Mendelian Inheritance

Mitochondrial inheritance

The human mitochondrion has a small circular genome of 16,569 bp, which is remarkably crowded. It is inherited only through the egg and the sperm does not contribute to the zygote population of mitochondria. There are relatively few human genetic diseases caused by mitochondrial mutations but, because of their maternal transmission, they have a very distinctive pattern of inheritance.

All the children of an affected female are affected but none of the children of an affected male inherit the disease (**Figure 2.15**).

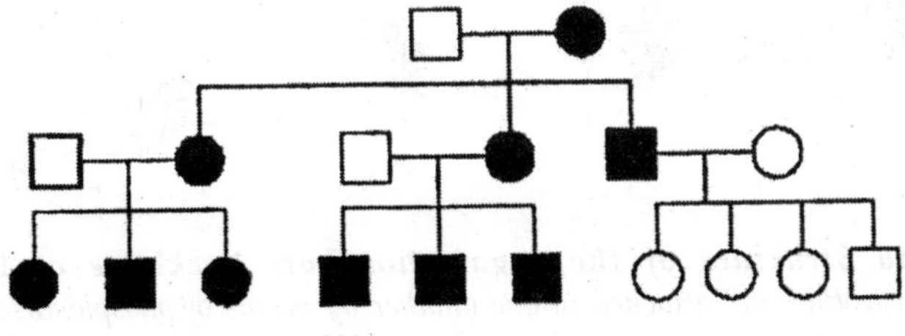

Figure 2.15 A mitochondrial inheritance pedigree

Uniparental disomy

It is sometimes possible that both copies of a single chromosome may be inherited from the same parent (with none of the corresponding chromosome from the other parent). Rare cases of cystic fibrosis (a common autosomal disease) have occurred in which one parent was a heterozygous carrier of the disease and the second parent had two wild type alleles. In such a case, the child would have received two copies of chromosome 7 from the carrier parent and no chromosome 7 from the unaffected parent, a condition known as uniparental disomy.

DNA

Genes control all characters and genes are made up of nucleic acids. There are two types of nucleic acids : deoxyribosomal nucleic acid (DNA) and ribonucleic acid (RNA). In this section we will discuss the structure of these nucleic acids and an introduction to the molecular techniques that have been and continue to be largely responsible for the advances in the field of immunogenetics.

Each nucleic acid consists of a sugar-phosphate backbone with bases attached to it (**Figure 2.16**). These nitrogenous bases are of two types, purines and pyrimidines. DNA consists of two purine bases, adenine (A) and guanine (G) and two-pyrimidine bases thymine (T) and cytosine (C) along with the sugar deoxyribose; in RNA uracil is replaced by thymine and the sugar present is ribose.

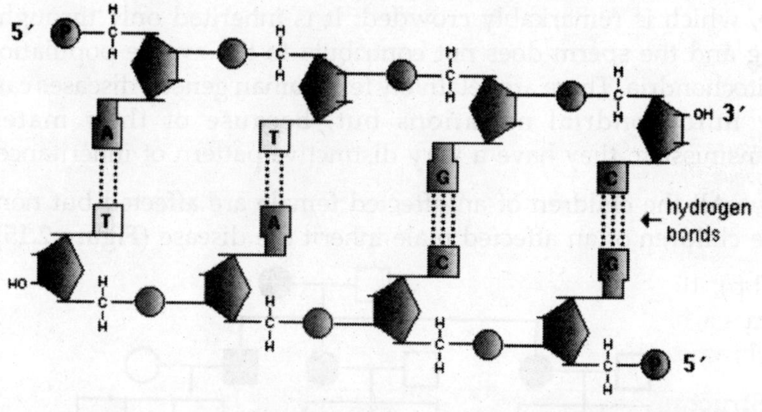

Figure 2.16 Structure of the sugar-phosphate backbone of DNA. *The deoxyribonucleotides are attached to one another by means of phosphodiester linkages between the 3′ - OH of one nucleotide and the 5′ -OH of the next nucleotide*

The nitrogenous bases are attached to the 1′ (one prime) position of each sugar and the phosphate links 3′ and 5′ hydroxyl groups of adjacent sugar moieties. A unit of purine or pyrimidine base attached to a sugar and a phosphate group is known as a nucleotide.

A molecule of DNA consists of two nucleotide chains, which are coiled clockwise around one another to form a double helix with 10 nucleotides per complete turn of DNA (**Figure 2.17**). The two chains run in opposite direction and are held together by hydrogen bonds between A and T and G and C of adjacent chains.

This base pairing is very specific, although erroneous combinations may rarely occur. Since A: T and G: C pairing is obligatory, the nucleotide sequences of parallel strands are complementary to each other and the ratio of A to T of G to C is 1: 1 (Chargaff's rule). However, variations exist between (A+T) / (G+C) ratio which ranges from 1: 4 to 1 in humans.

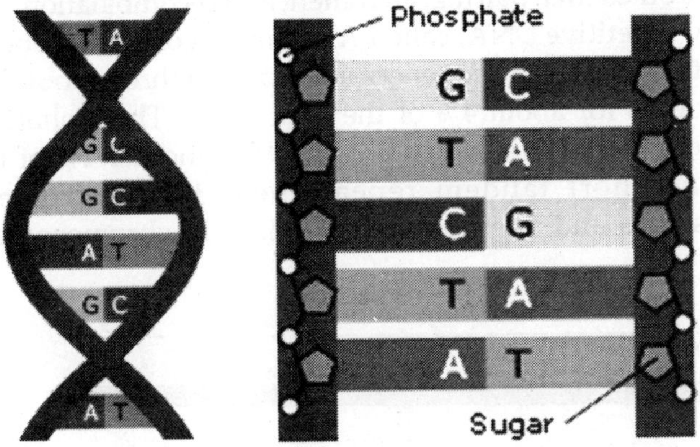

Figure 2.17 Structure of DNA showing base pairing between the two strands. *The ribbons indicate the sugar-phosphate backbones of the two strands; base pairing is shown between the two complementary strands. Together, the strands form a double helix.*

The length of DNA is expressed as the number of base pairs (bp); 1000bp is known as a kilobase (bp) and 10^6 bp is known as a mega base (mb). The total length of DNA in a half set of human chromosomes is 3000 mb (3 x 10^9 bp).

Structure of a gene

A gene consists of one or more segments of DNA whose sequence encodes a particular RNA molecule (**Figure 2.18**). By convention, sequences at the 5′ end of the DNA are said to be upstream from sequences located at the 3′ end. The DNA segments encoding a gene or its exons may be separated from each other by introns, or intervening sequences (IVS). The number of introns varies from gene to gene. For example, mouse beta globin gene has 2 introns.

Approximately 28-30,000 structural genes are encoded within the DNA. Each structural gene usually has only one copy in the haploid genome. If the average gene length is taken to be the total genetic material, 20kb only accounts for 1000 mega bases or about one-third of the total DNA. Much of the remainder consists of repetitive DNA sequences, which may be either moderately repetitive with several hundred copies or highly repetitive with many thousands of copies. These sequences are widely dispersed or occur in clusters. The moderately repetitive DNA includes some functional genes, which occur as multiple copies and include ribosomal RNAs (about 2000 copies) and the histone genes. In contrast, the highly repetitive DNA (perhaps 10% of the total DNA) is not transcribed and may have a role in chromosomal pairing, alignment and recombination. One family of highly repetitive DNA, called Alu family, consists of about 500,000 copies of 300 base pair sequences interspersed throughout the genome and accounts for about 4% of the total DNA. The alphoid family of highly repetitive DNA which accounts for another 4% of total DNA occurs as short tandem repeats near the centromere of all chromosomes and is especially abundant in chromosomes 1, 9, 16 and 4.

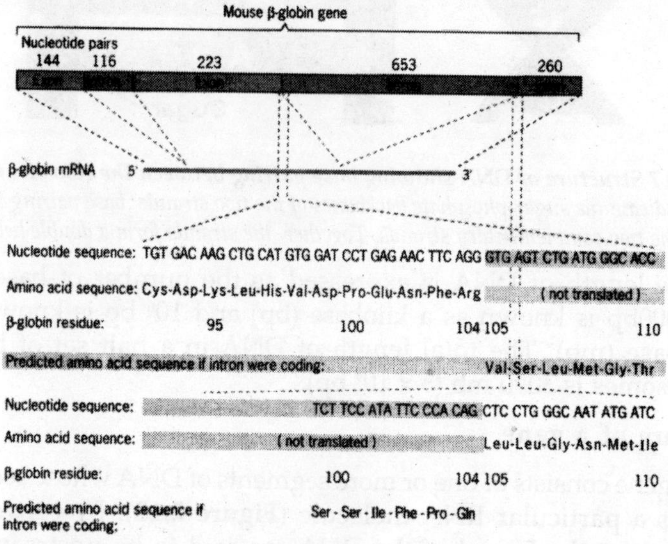

Figure 2.18 Structure of the mouse beta globin gene

RNA differs in structure from DNA in several respects:

1. The sugar is ribose rather then deoxyribose;
2. Uracil (U) replaces thymine (T);
3. RNA single stranded
4. Five types of RNA are found (**Table 2.5**)

The flow of information from DNA to RNA is known as transcription. The RNA copy of the DNA that is destined to form a protein molecule is known as messenger RNA or mRNA. Synthesis of the message is initiated from a promoter region in the DNA and primary RNA transcriptsconsist of number of coding and non-coding sequences (introns and exons). The mRNA leaves the nucleus and in the cytosol splicing takes place, which removes introns from the transcript (processing) to form a contiguous mRNA strand. Adenylic acid nucleotides are added to the terminal end of the mRNA molecule and the molecule is said to be polyadenylated. The mRNA is then transported to the cytoplasm and translated into protein.

Table 2.5: Types of RNA

Type	Location	Comments
Messenger RNA (mRNA)	Nucleus and cytoplasm	Variable size; base sequence complementary to transcribed DNA; about 4% of total cellular RNA; half-life 7-24 hours.
Transfer RNA (t-RNA)	Cytoplasm	Hairpin-loop shape; about 40 types; amino acid-specific; 73-93 nucleotides in each; about 10% of total cellular RNA with tens to hundreds of copies of genes for each tRNA species.
Ribosomal RNA (rRNA)	Ribosomes and nucleolus	About 80% of total cellular RNA; synthesized and stored in nucleolus.
Heterogenous RNA (hnRNA)	Nucleus	High molecular weight mRNA precursors
Small nuclear RNA (smRNA)	Nucleus	Six types (U1-U6) of 100-215 nucleotides each; involved in RNA splicing.

Translation occurs on the ribosomes. The first portion of the RNA molecule, the untranslated portion, is not a protein coding sequence. The amino acid sequence of a protein is determined by the ribonucleotide sequence of the mRNA, which was derived from the DNA sequence. For mRNA to be translated into the specific protein, interactions among translation initiation enzymes, cofactors, amino acid-charged transfer RNAs (tRNAs) and ribosome must take place. The coding region of the mRNA is dictated by a series of triplets of nucleotides called codons, each triplet specifying a particular amino acid. The relation of triplet codons to their respective amino acids is known as the **genetic code (Table 2.6)**. A series of codons that can be transcribed into a protein is known as **open reading frame**.

Table 2.6: Genetic code

First Base	Second Base U		C		A		G		Third Base
U	UUU	Phe	UCU	Ser	UAU	Tyr	UGU	Cys	U
	UUC	Phe	UCC	Ser	UAC	Tyr	UGC	Cys	C
	UUA	Leu	UCA	Ser	UAA	STOP	UGA	STOP	A
	UUG	Leu	UCG	Ser	UAG	STOP	UGG	Trp	G
C	CCC	Leu	CCU	Pro	CAU	His	CGU	Arg	U
	CUC	Leu	CCC	Pro	CAC	His	CGC	Arg	C
	CUA	Leu	CCA	Pro	CAA	Gln	CGA	Arg	A
	CUG	Leu	CCG	Pro	CAG	GLn	AGG	Arg	G
A	AUU	ILe	ACU	Thr	AAU	ASn	AGU	Ser	U
	AUG	ILe	ACC	Thr	AAC	ASn	AGC	Ser	C
	AUA	ILe	ACA	Thr	AAA	Lys	AGA	Arg	A
	AUG	Met	ACG	Thr	AAG	Lys	AGG	Arg	G
G	GUU	Val	GCU	Ala	GAU	Asp	GGU	Gly	U
	GUC	Val	GCC	Ala	GAC	Asp	GGC	Gly	G
	GUA	Val	GCA	Ala	GAA	Glu	GGA	Gly	A
	GUG	Val	GCG	Ala	GAG	Glu	GGG	Gly	G

DNA replication

Replication of DNA occurs during each cell division. The two strands separate at a number of points (upto 100 per chromosome) and each strand serves as a template on which the missing partner can be reconstructed by base pairing with free nucleotides. These nucleotides are bound together by action of the enzyme DNA polymerase and are hydrogen-bonded to the template strand. Replication proceeds in both directions from each initiation point until the two new strands of DNA are complete. This is known as semi conservative method of replication.

Protein Synthesis

The exact copy of a stretch of DNA including introns is

transcribed into an RNA molecule, within the nucleus. This precursor RNA molecule is then spliced into a mature RNA and enters the cytoplasm where it functions as mRNA, cDNA or ribosomal RNA.

In the eukaryotic system, **regulatory sequences** are present at the 5′ flanking region of genes. The first of these is located about 25 to 30 base pairs (bp) upstream from the RNA start site. It is a promoter site for RNA polymerase and consists of a consensus sequence TATAAA. It is also known as **TATA box** or the **Goldberg-Hogness** box. There is A CAAT box is found about 70-90 bp upstream of the TATAA box (**Figure 2.19**). The expression of given gene may also be stimulated by enhancer sequences located up to many thousands of base pairs from the actual starting point of the gene. The triplet ATG, which specifies the amino acid methionine, at the 5′ end of a gene is the initiation signal for protein synthesis on messenger RNA; at the 3′ end the codons TAA, TAG or TGA act as termination signal. Translation is initiated in the cytosol and amino acids are carried to the mRNA on tRNAs and transferred within the ribosome on to the growing chain of polypeptides. Thus, one ribosome moves along the length of mRNA for each polypeptide chain and the tRNA sequentially transfers the amino acid corresponding to the next nucleotide triplet. This ensures that the fidelity of the DNA sequence is finally reflected in the amino acid sequence. This also implies that the tRNA is specific for each amino acid and carries the recognition site on it (anticodon) for the codon.

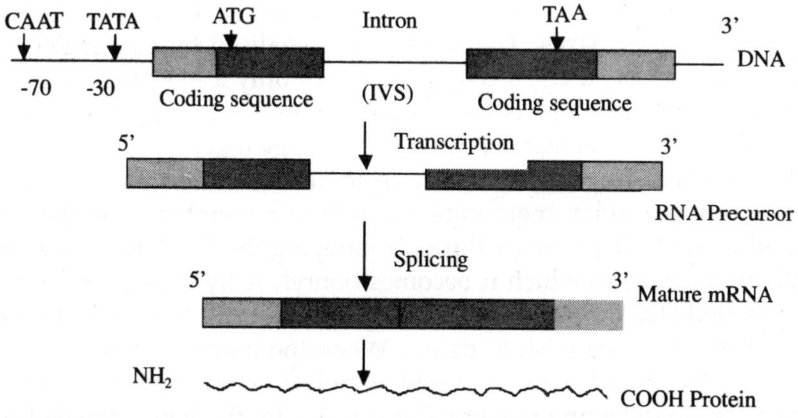

Figure 2.19 Model of a gene. *The regulatory TATA box and CAAT box are located upstream from the gene. In this gene the two exons are separated by an intron, the gene is transcribed to yield a precursor RNA molecule, which then spliced to yield a mature mRNA. The hatched areas represent protein-coding regions. The black areas in the RNA molecule represent 5′ and 3′ untranslated regions.*

The primary RNA transcript undergoes modifications. There are long stretches of DNA with no known function. Many sequences are repeated throughout the genome and are known as repeat sequences or **satellite DNA.**

Introduction to Techniques in Molecular Biology: After understanding the structures of the DNA, RNA and gene in the following section we will discuss the techniques used to study these molecules.

Southern Hybridization

In Southern hybridization, electrophoretically separated DNA fragments are transferred from an agarose gel to a nylon membrane and the fragment of interest is detected by hybridization probing. Besides its use in genome mapping, it has applications in other procedures that require identification of specific restriction fragments, for example to check the identity of a cloned fragment (**Figure 2.20**).

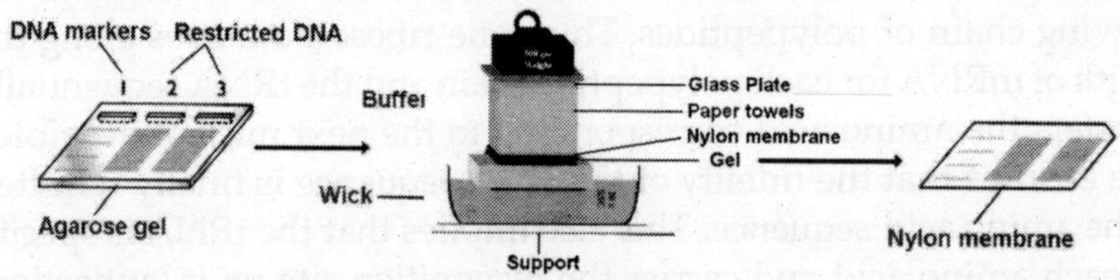

Figure 2.20 Electrophoresis and transfer of DNA for southern hybridisation

Agarose gel electrophoresis separates linear DNA molecules, according to their sizes. The bands are visualised by staining with ethidium bromide, a DNA dye. If there are only a few fragments in the digest, then each one is seen as a separate band. If there are more than about 50 fragments then the bands merge together into a smear and individual fragments cannot be identified. The fragments are transferred to a nylon membrane by placing a membrane on the gel and allowing buffer to soak through, carrying the DNA from the gel to the membrane to which it becomes bound. A hybridization probe is a labelled DNA molecule whose sequence is complementary to the target DNA that we wish to detect. When the membrane is exposed to the probe, it will bind or hybridize only to complementary DNA sequences on detecting the signal given out by the label attached to the probe, then its position is identified. The probe may be a synthetic oligonucleotide on a cloned DNA fragment. For detecting an RFLP, the probe is usually a cloned DNA fragment that spans the polymorphic restriction site.

To carry out the hybridization, the membrane is placed in a glass bottle with the labelled probe and some buffer, and the bottle is gently rotated for several hours so that the probe has adequate opportunity to hybridize to its target DNAs. The membrane is then washed to remove any probe that has not become hybridized and the signal from the label is detected. The detection of radioactive labelled probes is done by autoradiography (**Figure 2.21**). The visualized bands on autoradiograph are those that are complementary to the probe.

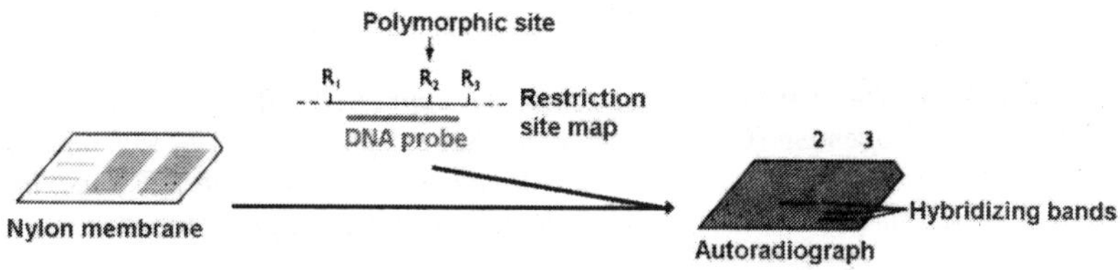

Figure 2.21 Development of autoradiograph after Southern Hybridization. *Lane 2 shows one band, while lane 3 shows 2 bands indicating that the DNA has a restriction site that was cut by the enzyme used in the experiment.*

Northern Hybridization

This is a technique, which is identical to southern hybridization except that RNA target sequences are detected using DNA probes. It is a useful method to assess the efficiency of gene transcription in a given tissue or cell.

RNA is isolated, electrophoresed in an agarose gel and transferred to a suitable support medium like nylon as is done for Southern hybridization. Because the DNA blot was called Southern (after the scientist who described it) a blot of RNA was named a Northern blot. (By analogy, a blot of protein is called a Western blot). The membrane is hybridized using a labelled probe and developed as an autoradirograph.

A Northern analysis gives information on how abundant a gene transcript is. The more RNA a band contains, the more probe it will bind and the darker the band will be on the film, which can be quantified by measuring the amount of light it absorbs in a densitometer (**Figure 2.22**).

Polymerase Chain Reaction

Polymerase chain reaction (PCR) involves exponential amplification of a selected region of a DNA molecule (**Figure 2.23).**

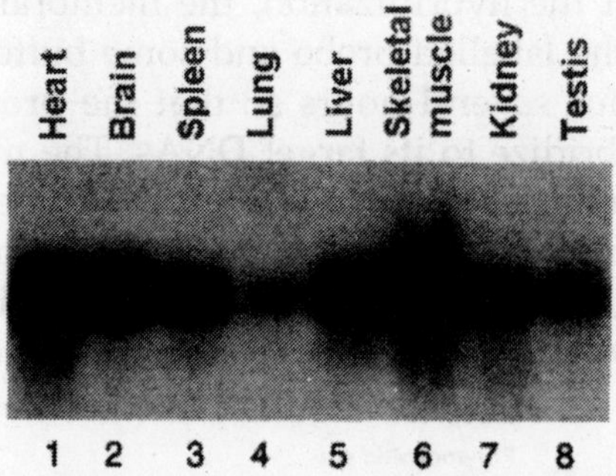

Figure 2.22 A Northern blot to assess relative expression of glyceraldehydes–3-phosphate dehydrogenase (G3PDH) gene. *mRNA extracted from different rat tissues has been electrophorased and probed with radrolabelled G3PDH probe. The maximum expression is seen in heart and skeletal muscle with least expression in lung.*

The essential first step is to design and synthesize primers, which span the DNA sequence to be amplified. Each primer should be complementary to the end regions of the gene segment to be amplified.

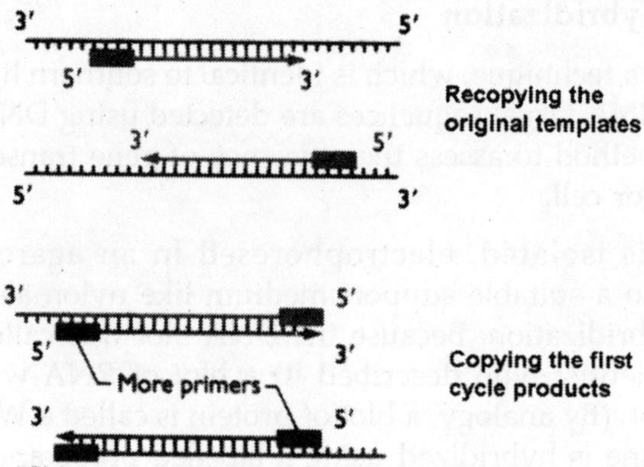

Figure 2.23 Polymerase chain reaction

The reaction involves target DNA molecules, which may be present in extremely low quantity, nucleotide primers and a thermostable DNA polymerase *Taq polymerase,* that is resistant to denaturation by heat treatment. *Taq polymerase* was originally isolated from the bacterium *Thermus aquaticus,* which lives in hot springs. The synthesis of new complementary polynucleotides occurs in the 5′ to 3′ direction by copying the template DNA in the 3′ to 5′ direction. Because the Taq polymerase is thermostable, the reaction mixture can be heated to 90°C without destroying the enzyme activity. At this temperature, the new strands detach from the template DNA. When the mixture is cooled down again, more primers anneal to the template DNA and also to the new strands and the Taq polymerase carries out a second cycle of DNA synthesis.

The reaction is performed for 30-40 cycles before the enzyme, primers or nucleotides are used up. A single starting molecule is therefore amplified into tens of millions of identical fragments, representing a few micrograms of DNA.

Regions of 5 kb and more can be amplified without difficulty, and longer amplifications - up to 40kb - are possible using modifications to the standard technique. The technique has many applications in molecular biology, such as in DNA mapping, DNA sequencing and molecular phylogenetics. A modified version of PCR can be used to amplify DNA copies of RNA molecules. PCR is also used extensively in specialised areas such as forensic science and clinical diagnosis, where its ability to work with very few starting molecules is particularly useful.

Oligonucleotide Primers

One of the most important technological advances in molecular biology is the automated synthesis of oligodeoxyribonucleotide chains.

Oligonucleotides can be used either as short probes or primers for the polymerase chain reaction (PCR). These single strands of deoxyribonucleotides are usually between 6 and 100 bases long, depending upon their application. They are synthesized on programmable machines that chemically bond successive nucleotides together to produce the desired nucleic acid.

DNA Restriction analysis

DNA can be cleaved by restriction enzymes, which recognize short sequences (usually 4 to 8 bp palindromes). There are several hundred restriction enzymes, which recognize 150 specific nucleotide sequences in double stranded DNA. List of some of the restriction enzymes along with their recognition site is shown in **Table 2.7**.

In human DNA, which is approximately 3 x 10^9 bp long, a restriction enzyme that recognizes and cuts at a 6 base pair recognition sequence, cuts DNA into several million fragments. The movement of the DNA fragments in an agarose gel matrix are approximately proportional to the log 10 of their length in base pairs. These fragments can be visualized under ultraviolet light with the help of ethidium bromide staining (**Figure 2.24**).

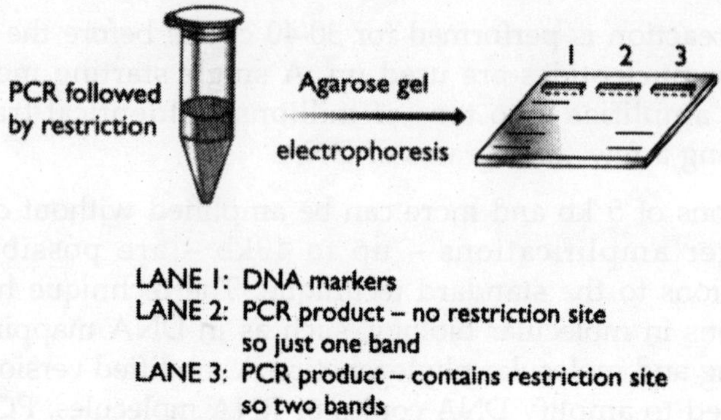

Figure 2.24 Restriction digestion of PCR product

Pulsed- Field Gel Electrophoresis

The Southern blot technique suffers from the limitation that fragments larger than 20 kb are poorly resolved on conventional agarose gels. In a variation of this method, pulse field electrophoresis (PFGE), resolution is increased by using pulses of current for electrophoretic separation instead of constant current. The DNA molecules align in a linear fashion within the field when pulses are applied. In this method current is applied to the gel at different angles and the polarity of the field is biased to favour DNA mobility in one direction as pulses of current are applied. The size range in which fragments resolve using FIGE depends on the placement of the electrodes and the duration of the pulses. Fragments as large as 7 million base pairs have been separated in this way. Large fragment DNA analysis uses restriction enzymes that recognize larger recognition sites (eg 8 - bp sequences) which theoretically cleaves a random DNA strand once every 65,536 bp.

RNAse Protection Analysis

This method involves protection of double stranded hybrids from the action of nucleases and can be used to quantitate RNA. A DNA fragment complementary to the specific RNA in question is radio-labelled, denatured and mixed with total or poly A-enriched RNA under conditions that favour RNA-DNA hybridization. The enzyme

Table 2.7: A list of restriction enzymes and their recognition site

	Enzyme name	Organism in which enzyme is found	Recognition sequence and position of cut
Enzymes with 6 bp recognition sequences	Bam HI	Bacillus amyloliquefaciens H	5′ G↓G A T C C 3′
			3′ C C T A G↑G 5′
	BgII	Bacillus globigi	5′ A↓G A T C T 3′
			3′ T C T A G↑A 5′
	Eco RI	E. coli RY13	5′ G↓A A T T C 3′
			3′ C T T A A↑G 5′
	HindIII	Haemophilus influenzae Rd	5′ A↓A G C T T 3′
			3′ T T C G A↑A 5′
	PstI	Providencia stuartii	5′ C T G C A↓G 3′
			3′ G↑A C G T C 5′
	SalI	Streptomyces albus	5′ G↓T C G A C 3′
			3′ C A G C T↑G 5′
	SmaI	Serratia marcescens	5′ C C C↓G G G 3′
			3′ G G G↑C C C 5′

Enzymes with 4-bp recognition sequences	HaeIII	Haemophilus egyptius	5′ GG↓CC 3′ 3′ CC↑GG 5′
	HhaI	Haemophilus hemolyticus	5′ GCG↓C 3′ 3′ C↑GCG 5′
	HpaII	Haemophilus parainfluenzae	5′ C↓CGG 3′ 3′ GGC↑C 5′
	Sau3A	Staphylococcus aureus 3A	↓GATC CTAG↑
Enzymes with 8-bp recognition sequences	NotI	Nocardia otitidis-caviarum	GC↓GG CCGC CGCCGG↑CG
Enzymes with recognition sequences that are not symmetrical	BstXI	Bacillus stearothermophilus	CCANNNNN↓NTGG GGTN↑NNNNNACC

S1 nuclease that digests only single stranded DNA and RNA molecules is added; it chops up all the single strands leaving hybrid molecules intact. These complexes are separated on polyacrylamide gel electrophoresis. Because RNA-DNA annealing conditions are selective, S1 nuclease protection analysis is very sensitive and can be used to detect one mRNA molecule in 100 cells if large amounts of RNA are available. This method is useful for low abundance transcripts. A variation of this method uses RNA-RNA hybridization condition.

Western blot analysis

Western blot analysis is used to detect proteins. Proteins are subjected to electrophoretic separation on sodium dodecyl sulphate polyacrylamide gel (SDS-PAGE) and transferred to a nitrocellulose membrane using high voltage electrophoresis. The membrane is sequentially exposed to specific antibodies and enzyme or luminescent linked secondary antibodies. The protein bands recognised by the antibodies are visualised by the colour reactions or by auto radiographic methods.

DNA sequencing

Nucleotide base order can be determined by DNA sequencing. Historically two methods have been used. Sanger in 1977 developed dideoxy termination sequencing procedure and Gilbert developed chemical degradation sequencing protocol. In the Maxim - Gilbert technique, single stranded DNA is subjected to four different reactions which cleave randomly at one of the four sites. The samples are then subjected to polyacrylamide gel electrophoresis, which separates the strands by size.

DNA is synthesized from deoxyribonucleoside triphosphates (dNTPs), which have an OH group on the 3′-carbon atom (**Figure 2.25a**). During DNA synthesis, two phosphate groups on the 5′-carbon atom of a dNTP are removed, and a phosphodiester bond is formed between the remaining 5′-phosphate group of the dNTP and the 3′-OH group. (When of the last nucleotide on the growing DNA chain.) In the Sanger method, a special nucleotide, called a dideoxyribonucleoside triphosphate (ddNTP; **Figure 2.25b**), is used as substrate. The ddNTPs are identical with dNTPs, except that they lack a 3′-OH group. Hence, when they are incorporated into a growing DNA chain through the phosphate groups on their 5′ ends, further nucleotide cannot be added, because there is no 3′ -OH group to form a phosphodiester bond with an incoming nucleotide. Thus, ddNTPs terminate DNA synthesis.

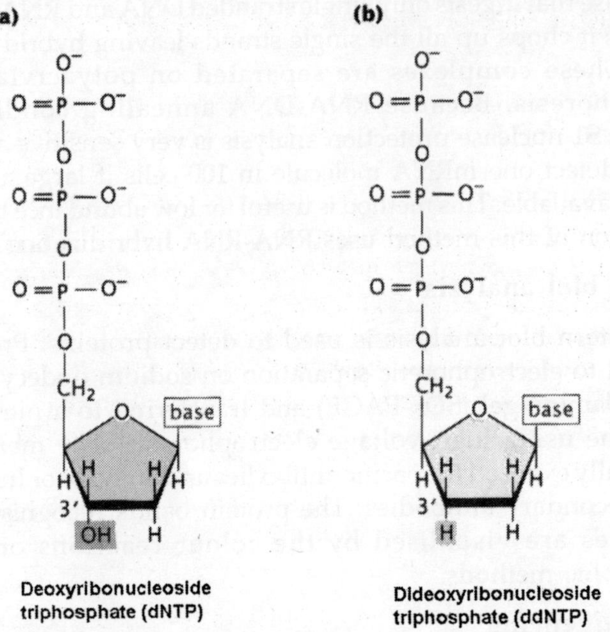

Figure 2.25 The dideoxy sequencing reaction requires a special substrate for DNA synthesis. *(a) Structure of deoxyribonucleoside triphosphate, the normal substrate for DNA synthesis. (b) Structure of dideoxyribonucleoside triphosphate, which lacks an OH group on the 3' carbon atom.*

A DNA fragment to be sequenced is amplified by PCR or by cloning in bacteria to have adequate DNA for this reaction. The target DNA is added to four reaction tubes, to which are added:

1. Many copies of a primer that is complementary to one end of the target DNA strand;
2. All four deoxyribonucleoside triphosphates (dCTP, dATP, dGTP, and dTTP), the normal precursors of DNA synthesis;
3. A small amount of one of the four types of dideoxyribonucleoside triphosphates (ddCTP, ddATP, ddGTP, or ddTTP), which will terminate DNA synthesis as soon as it is incorporated into any growing chain (each of the four tubes receive different ddNTP);
4. DNA polymerase.

Either the primer or one of the dNTPs is radioactively or chemically labelled so that newly produced DNA can be detected.

DNA polymerase enzyme carries out DNA synthesis and each

of the single strands of target DNA serves as a template for DNA synthesis. The primer pairs to its complementary sequence at one end of each template strand, providing a 3′ -OH group for the initiation of DNA synthesis. DNA polymerase elongates a new strand of DNA from this primer by using the target DNA strand as a template. As an example let us consider the reaction tube with reaction ddATP. When DNA polymerase encounters a T on the template strand, it uses at random either a dATP or a ddATP to introduce an A in the newly synthesized strand. Because there is more dATP than ddATP in the reaction mixture, dATP is incorporated more often, allowing DNA synthesis to continue. Occasionally, however, ddATP is incorporated into the strand and synthesis terminates. The incorporation of ddA into the new strand occurs randomly at different positions in different copies, producing a set of DNA chains of different length (**Figure 2.26**), each ending in adenine. Similar reactions take place in the other three tubes having ddCTP, ddGTP and ddTTP, producing chains terminating in cytosine, guanine and thymidine respectively. At the end of the polymerization reactions, the entire DNA is denatured, and the single-strand products of each reaction separated by gel electrophoresis in adjacent lanes.

The DNA strands differing in length by only a single nucleotide are easily distinguished on the autoradiograph. The DNA sequence can be easily read and is the shortest part of the procedure. In **Figure 2.26** the band closest to the bottom of the gel is from ddGTP reaction, indicating that the first nucleotide synthesized had guanine (G). The next band from the ddATP reaction; and so on. In this manner, the sequence can be read from the bottom to the top of the gel, with the nucleotides near the bottom corresponding to the 5′ end of the newly synthesized DNA strand and those near the top corresponding to the 3′ end. It should be kept in mind that the obtained sequence is not of the target DNA but of its complementary strand.

If the sequence of the target DNA is unknown, the target DNA is cloned in a vector that contains sequences recognized by a common primer (called universal sequencing primer sites) on either side of the inserted sequence of interest. The target DNA is then isolated from the vector and sequencing reaction is set up using the universal sequencing primers (**Figure 2.27**).

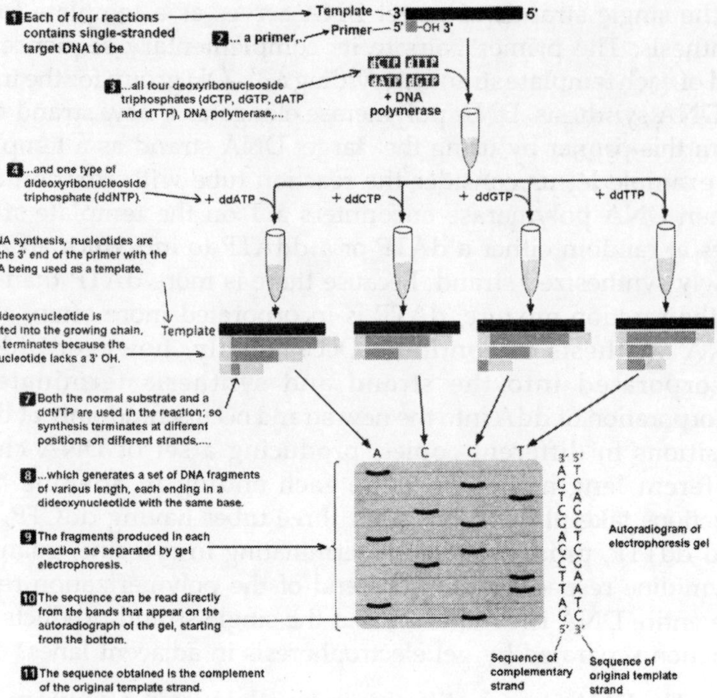

Figure 2.26 The dideoxy method of DNA sequencing is based on the termination of DNA synthesis.

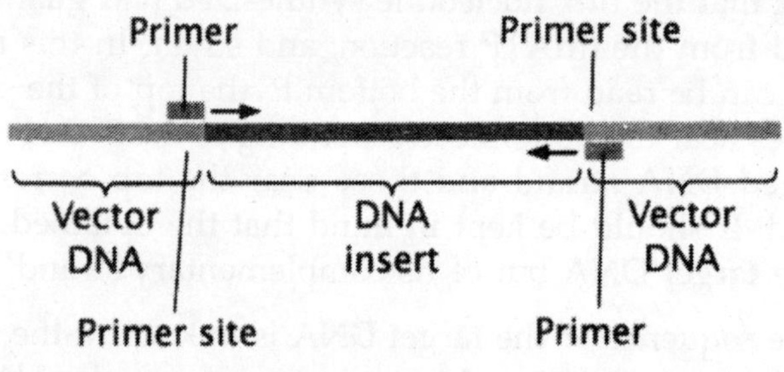

Figure 2.27 Sites recognized by sequencing primers *are added to the target DNA by cloning the DNA in a vector that contains universal sequencing primer sites on either side of the site where the target DNA will be inserted.*

Today sequencing is done on automated machines that use fluorescent dyes and laser scanners to detect the signals. By this method, thousands of base pairs can be sequenced in few hours. In the dideoxy reaction, a different coloured dye is used for each type of dideoxynucleotide. The four sequencing reactions take place in the same tube and are electrophoresed in the same well, the resulting bands being scored by the colour of the terminal nucleotide. The recent sequencing machines have gel-containing capillary tubes to run the electrophoresis and the products are passed through a laser beam and detector. As the fragments pass the laser, their fluorescent dyes are activated and an optical scanner detects the resulting fluorescence. Each coloured dye emits fluorescence of a characteristic wavelength, which is read by the optical scanner. The information is fed into a computer for interpretation and the results are printed out as a set of peaks on a graph (**Figure 2.28**). Automated sequencing machines may contain 96 or more capillary tubes, allowing from 50,000 to 60,000 bp of sequence to be read in a few hours.

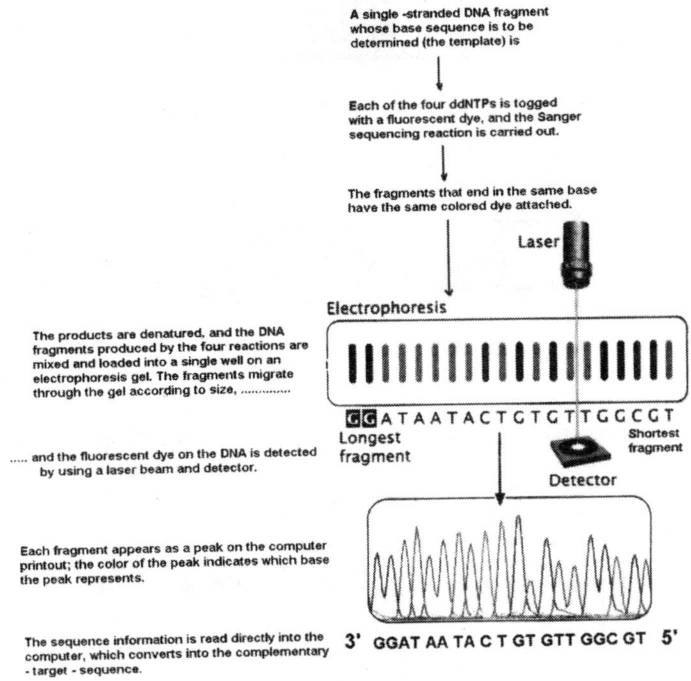

Figure 2.28 Automated dideoxy sequencing method

Molecular Cloning

Molecular cloning is the technique of isolation and amplification of defined DNA fragments. It involves the manipulation of DNA with enzymes followed by screening with probes to isolate the desired nucleic acid sequence. PCR technology has significantly simplified this process since the required segment can be amplified in abundance by PCR and tedious steps involved in obtaining the DNA of interest from the genomic DNA are bypassed.

The first step in cloning is cleavage of DNA (genomic) into multiple fragments. A critical enzyme used in molecular cloning is **DNA ligase**. This enzyme reseals the cohesive ends of two DNA fragments as shown in the **Figure-2. 29.**

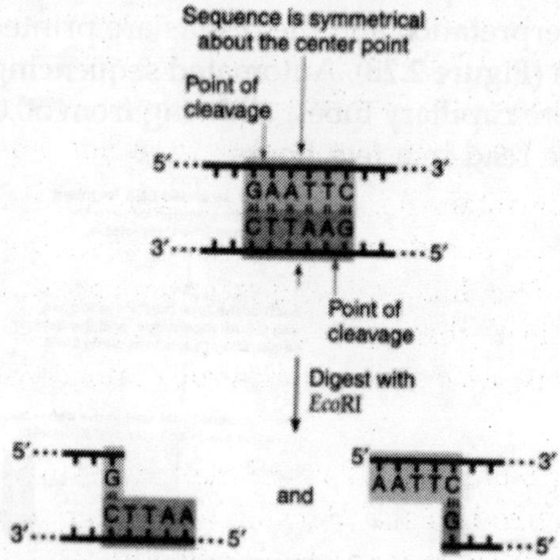

Digestion of DNA with Eco RI

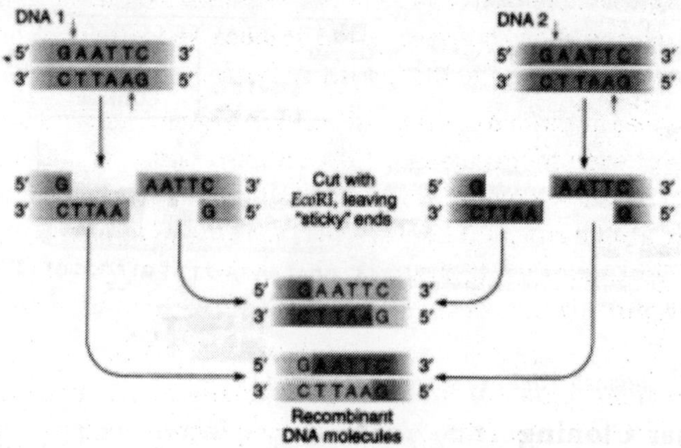

***Figure 2.29* Recombinant DNA molecules.** *The construction of recombinant DNA molecule invitro. DNA isolated from two different species are cleaved with a restriction enzyme, mixed under annealing conditions, and covalently joined by treatment with DNA ligase. The DNA molecules can be obtained from any species animal, plant or microbe. The digestion of DNA with the common cloning vectors.*

For example, the genome of polyoma virus, a circular double stranded 5292 bp long DNA virus, contains a single Bam III recognition site. The restriction enzyme Bam III linearizes the molecule; addition of ligase recircularizes the molecule.

DNA fragments to be cloned are inserted into constructs called **vectors**. The major types of vectors are **plasmids, bacteriophages, cosmids** and **yeast artificial chromosome** (YAC) as shown **in the Figure 2.30**. Once the recombinant molecules are formed, the production of many copies or clones of these molecules is carried out.

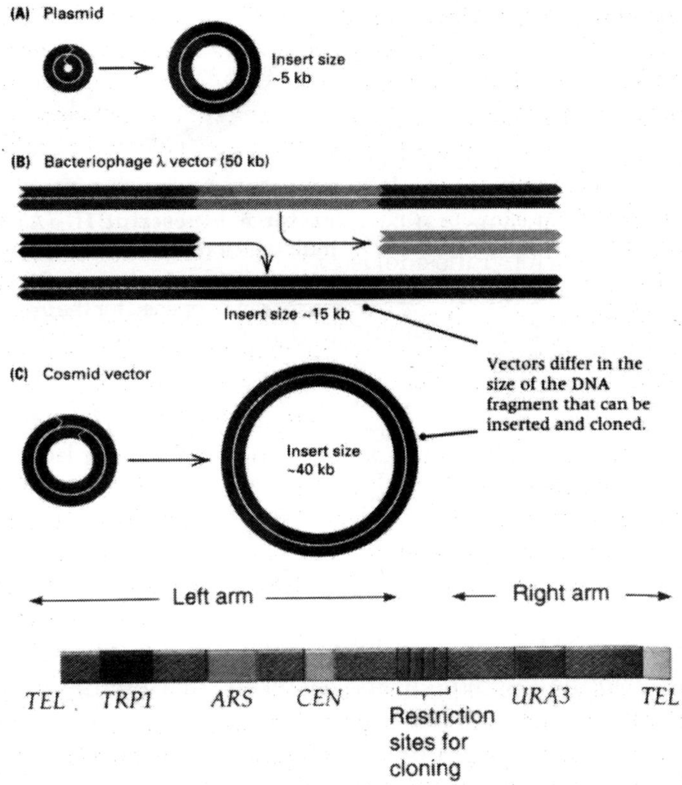

Figure 2. 30 Various vectors used in cloning. *(A) Plasmid vectors are ideal for cloning relatively small fragments of DNA (B) Bacteriophage λ vectors contain convenient restriction sites for removing the middle / section of the phage and replacing it with the DNA of interest. (C) Cosmid vectors are useful for cloning DNA fragments up to about 40kb, they can replicate as plasmids but contain the cohesive ends of phage λ and so can be packaged in phage particles.*

Cloning vector has three essential components (1) an origin of replication (2) a dominant selectable marker gene, usually a gene that confers drug resistance to the host cell and (3) at least one unique restriction endonuclease cleavage site (**Figure 2.31**).

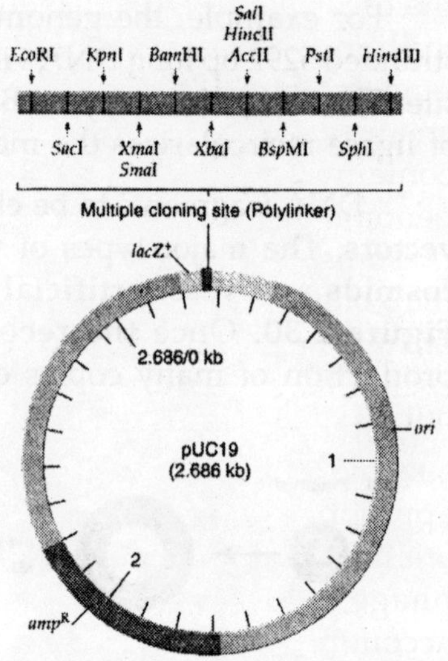

Figure **2.31 Diagram of the cloning vector ñ bluescript II.** (A) The multiple cloning site containing the unique restriction sites at which the vector can be opened for the insertion of DNA fragments. (B) It contains a plasmid origin of replication, an amplicillin resistance gene, a multiple cloning site (polylinker) within a fragment of the lac Z gene from *E coli* and a bacteriophage origin of replication.

The vector is ligated with the same restriction enzyme that is used to cut the genomic DNA. If the foreign DNA is ligated into a restriction site that disrupts the drug resistant gene of the vector, it can be used as a tool to select for those clones into which the foreign gene has been introduced. Lac Z gene is also used as a marker for insertion. X-gal (5-bromo-4 chloro-3 indoly1 – B-D-galactopyranoside) is an artificial substrate for the Lac Z encoded enzyme. Bacteria containing the vector produce blue colonies when grown on agar containing X –gal and a Lac Z gene promoter, isopropylthio-B-D galactopyranoside. Plasmids that have incorporated inserts are white. A combination of drug resistance and Lac Z expression can be used to select for bacteria that have incorporated a plasmid and have an insert.

For larger DNA fragments, bacteriophage has been used as a vector. Bacteriophages, which are bacterial viruses, have been modified to function as vectors by removing sections of their genome that are not essential for autonomous replication. Foreign DNA fragments up to 20 kb can be cloned into these vectors. E. coli is the host of choice for many of the manipulations performed on cloned DNAs. Thus some of the most useful cloning vectors are shuttle vectors that can replicate in both E. coli and another species, such as eukaryotic organism.

A genomic library is a collection of recombinant molecules that theoretically represents every DNA sequence within the genome of the organism being studied. One million recombinant clones of random genomic fragments with an average length of 15kb should contain 15 billion base pairs of genomic sequences and would presumably be representative of the 3 billion base pair mammalian or human genome. To construct such a library, genomic DNA is digested with a restriction enzyme to yield fragments of the appropriate size. The fragments are mixed with a vector (usually phage, because they accommodate large fragments) that has been cut with the same restriction enzyme. Ligase is added to join the vector with the inserts. After packaging *in-vitro* with phage coat proteins the infectious recombinant phages are mixed with susceptible bacteria and plated on agar. Bacteria infected with the phage (300 phages per cell) are lysed after phage proteins are produced and clear areas known as plaques form on the agar surface.

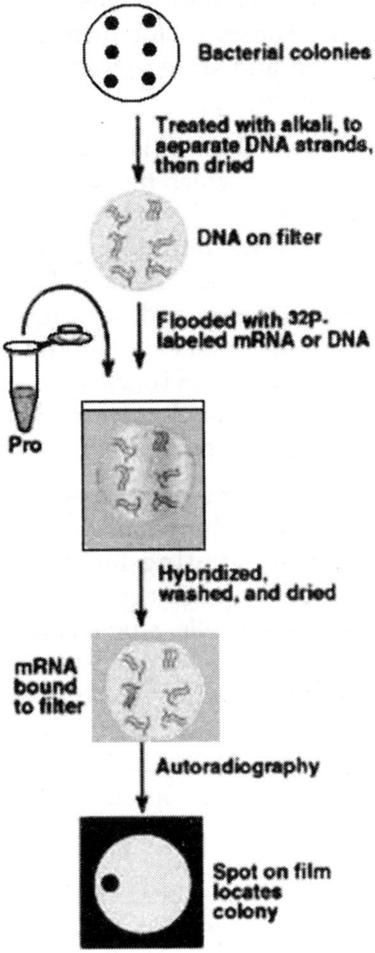

Figure 2.32 Screening DNA libraries by colony hybridization. A radioactive cDNA is employed as a hybridization probe.

When bacterial plasmids are used as vectors, ligated recombinants are transformed into a competent bacterial host that is plated on agar. An adequate number of colonies are screened by southern hybridization using a probe to detect the desired recombinant DNA clone (**Figure 2.32**). Colonies or plaques are transferred to membranes, the filters are treated with alkali to denature the DNA and filters are hybridized with probe of interest. Positive colonies are identified by autoradiography and matched to colonies on the agar plates. By this method, a single recombinant molecule can be identified from among the millions of host genomic fragments. Once the recombinant colony, containing the desired DNA fragment is identified, it can be grown in large quantities for further analysis. Such DNA libraries can be amplified and stored indefinitely. They can be screened repeatedly with different probe.

cDNA cloning is a powerful tool to study cell specific gene expression in which the starting sample is mRNA instead of genomic DNA. mRNA is first reverse transcribed to get single stranded DNA which is made into double stranded DNA using DNA polymerase. This is then cloned into a vector. In this approach only the transcribed genes are cloned and, therefore, the number of colonies are greatly reduced and makes screening easier. These libraries are useful in providing amplified copies of specific sequences, for example, a cDNA library from reticulocytes will contain many copies of globin cDNA, whereas this gene occurs only twice in a genomic library.

cDNA libraries have been useful for identifying transcripts that are differentially regulated in related cell types in the so-called **subtraction cloning technique.** For example, to identify novel transcripts that are expressed in mitogenically stimulated cells but not in resting cells, single strand cDNA is prepared from the latter. RNA -DNA hybrids representing sequences common to both cell types are removed by several separation techniques. This enhances the concentration of novel sequences for the mitogenically stimulated cell population. The remaining single strand DNA is used to generate a subtraction cDNA library.

Cloning large DNA fragments

Vectors have been developed for use in cloning fragments larger than 20kb. A **cosmid vector** has the part of the plasmid genome that allows autonomous replication and the phage sequences essential for packaging into the phage virion. Up to 50 kb of DNA can be inserted into the cosmid vector. Lambda phage and cosmid libraries are useful for chromosomal walking. In this technique the end of one clone is used as a probe for rescreening libraries to select recombinants that contain adjacent chromosomal DNA sequences. Large segments of the mouse and human MHC loci have been isolated in this manner.

Vectors that are used to clone fragments larger then 100 kb such as mini chromosomes in yeast have centromeric and telocentric regions of the yeast chromosomes combined with large fragments of foreign DNA (**Figure 2.33**) to sequence the total genome. The international effort to sequence the total human genome was known as the Human Genome Project. Having achieved its target, presently maps are being

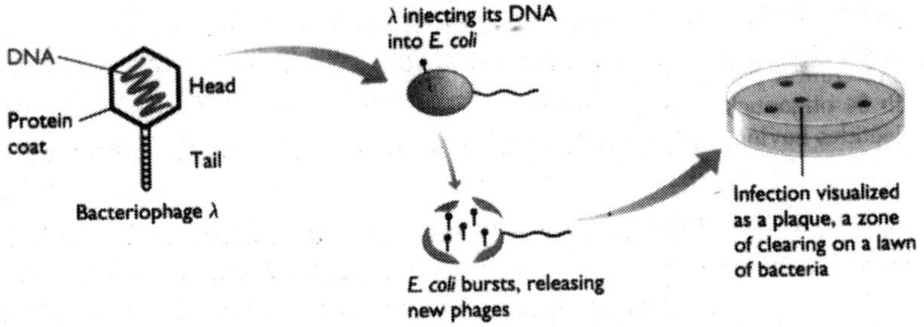

Figure 2.33 ***Vectors for cloning DNA fragments upto 44kb in length.*** *Bacteriophage can accommodate upto 52kb of DNA. Bacteiophage DNA injected into E coli, E coli bursts releasing new phages.*

constructed to locate genes responsible for monogenic and multifactorial disorders.

The genetic maps are constructed from recombination frequencies, using molecular markers such as restriction fragment length polymorphisms (RFLP's). Cytogenetic maps are maps which are based on the banding patterns of chromosomes. Physical maps often contain the locations of overlapping genomic clones or contigs and unique nucleotide sequences called sequence tagged sites or STS. Markers that are mapped both genetically and physically are called anchor markers. For this purpose PCR is used to recognize a particular sequence wherever it occurs. The different types of maps can all be based on STSs, allowing maps to be aligned. A special form of STS is based on the microsatellites used as markers (**Figure 2.34**). The definitive genetic map of the human genome has now been produced. Physical maps locate features to a particular part of the DNA.

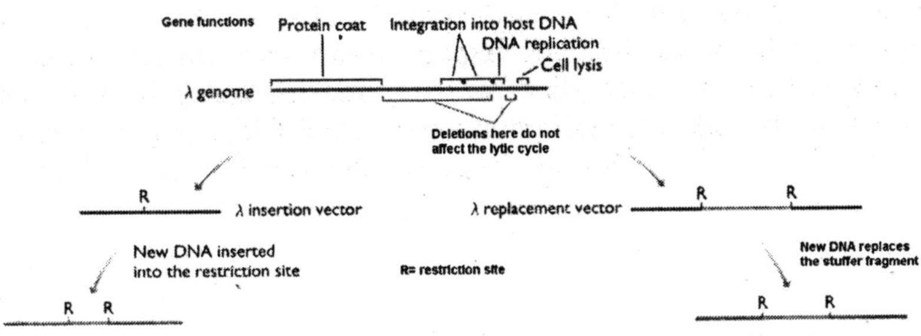

Figure 2.34 ***Genomic mapping using sequence tagged site (STS).*** *An STS marker is a DNA sequence, present once per haploid genome, that can be amplified by polymerase chain reaction.*

Clone maps

Clone maps use the overlap between clones in a genomic library to construct groups of clones called contigs, where the spatial relationship of each clone is defined with respect to the other clone in the contig. Comprehensive clone maps are constructed using YAC (yeast artificial chromosome) as vectors because of the large insert size they can carry.

Radiation hybrid (RH) maps

In RH mapping, a human cell line is irradiated with X-rays before fusion to hamster cells to create somatic cell hybrids. Radiation causes fragmentation of human chromosomes such that they cannot remain independently in the hybrid unless they become incorporated into the hamster chromosomes. Each hybrid line retains about 20% of the donor human genome in fragments of about 10 Mb. A panel of lines are more frequent than would be expected by chance they are linked. The distance between markers is estimated from the frequency of co-retention.

RH maps are valuable as they provide an independent method of mapping STS sites. Because it does not rely on the generation of clones it is not subjected to the same systematic errors, which may affect clone maps. Moreover, RH maps can detect linkages between STSs over a much longer distance than can be detected by the overlap of clones. They, therefore, provide a framework to link clone contigs together.

Long-range restriction maps

Restriction maps with long sequences is an important form of physical map for DNA sequences of up to a few tens of kilobases. The landmarks mapped are the sequences at which different enzymes cut the DNA molecule. There are two reasons why conventional methods of producing restriction maps cannot be used to generate longer maps, which might extend over hundreds of kilobases or even megabases.

Commonly used enzymes, such as EcoRI or BamHI, cut at 6-bp target sequences. If random sequence DNA were digested with such an enzyme, the size would be very large and practically difficult to handle.

Rare cutting enzymes (**Table 2.8**), cut DNA less frequently than normal enzymes for the following reasons.

Table 2.8: Some examples of enzymes that rarely cut human DNA

Mlu I	ACGCGT
Not I	GCGGCCGC
Nru I	TCGCGA
Sal I	GTCGAC
Sfi I	GGCCNNNNNGGCC
Xho I	CTCGAG

- Some enzymes such as Not I have an 8-bp recognition target. In random sequence DNA this will only occur every 65 kb.
- Human DNA contains the dinucleotide CpG which is the target for DNA methylation and which prevents restriction enzymes from cutting the DNA.
- The restriction enzyme Not I cuts human DNA on an average every 500 kb and its target sites provide landmarks that are conveniently spaced for the analysis of human chromosomal regions. Other enzymes, such as Nru I and Sfi I also produce fragments that are large enough to be useful landmarks for chromosomal maps. Note that enzymes such as Xho I and Sal I are comparatively rare cutters of human chromosomal DNA, even though they are commonly used in the construction of conventional restriction maps. The large DNA molecules are separated by **pulsed field gel electrophoresis** (PFGE). Long-range restriction maps are important for positional cloning, since they provide the framework to locate the position of clones derived from techniques such as chromosome walking.

An integrated physical and genetic map of the human genome

STSs are used as common markers to align genetic, RH and YAC-based clone maps of the human genome. Genetic maps can detect linkage between markers over distances of 30 Mb, while RH mapping can detect linkage over distances of 10 Mb. As a consequence, comprehensive genetic and RH maps of the genome require only a few thousand markers; however, the resolution of such maps is low. In contrast, YAC-based clone maps which detect linkage over 1 Mb base range require a large number of markers to provide adequate coverage of the genome but provide high fine-structure resolution.

Clone-based maps suffer from two particular problems; the gaps between the contigs and detection of false linkage between markers as a result of chimeric clones. RH and genetic maps provide a framework for clone-based maps and allow adjacent clone contigs to be positioned with respect to each other thus trapping errors that result from chimeric clones. Work is in progress to construct an integrated map that aligns the Genethon genetic map with an RH map and an STS-content map based on YAC clones. This map contained 24568 STS markers in total; with 10850 STSs mapped by clone content, 14665 RH-mapped STSs and 5254 genetic markers form the Genethon map.

DNA footprinting

Many important DNA sequences serve as binding sites for proteins, such as the consensus sequences in promoter regions, which are binding sites for transcription factors. A technique called **DNA footprinting** is used to determine DNA sequences which are bound by such proteins. In a typical DNA-footprinting experiment, purified DNA fragments are labelled at one end with a radioactive isotope of phosphorus, ^{32}P. An enzyme or chemical is used to cleave the DNA randomly into sub-fragments, which are then denatured and separated by gel electrophoresis. The positions of the sub-fragments are visualized with autoradiography. This procedure is carried out both in the presence and in the absence of a

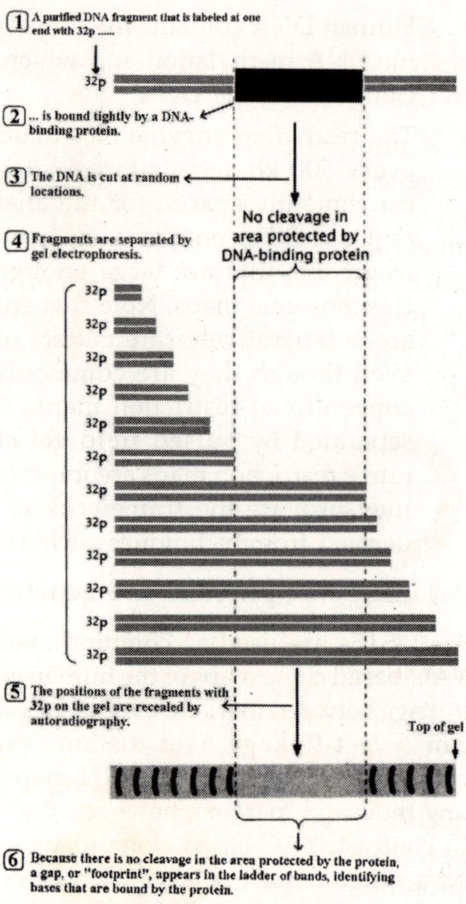

Figure 2.35 The steps in DNA foot printing: *used to determine DNA sequences which are bound by proteins.*

particular DNA-binding protein. When the protein is absent, cleavage is random along the DNA, producing a continuous "ladder" of bands on the auto-radiograph (**Figure 2.35**). When the protein is present, it binds to its target specific nucleotide sequence and protects their phosphodiester bonds from cleavage. Therefore, there is no cleavage in the area protected by the protein, and no labelled fragments terminating in this region appear on the autoradiograph. Their omission leaves a gap, or "footprint" on the ladder of bands (**Figure 2.35**) and the position of the footprint identifies those nucleotides bound tightly by the protein.

Mutagenesis

A powerful way to study gene function is to create mutations at specific locations by a process called **site-directed mutagenesis** and then to study the effects of these mutations on the organism.

A number of strategies have been developed for site-directed mutagenesis. One such strategy is to eliminate a short sequence of nucleotides with restriction enzymes and replace it with a short, synthetic oligonulceotide that contains the desired mutated sequence (**Figure 2.36**). The success of this method depends on the availability of restriction sites flanking the sequence to be replaced.

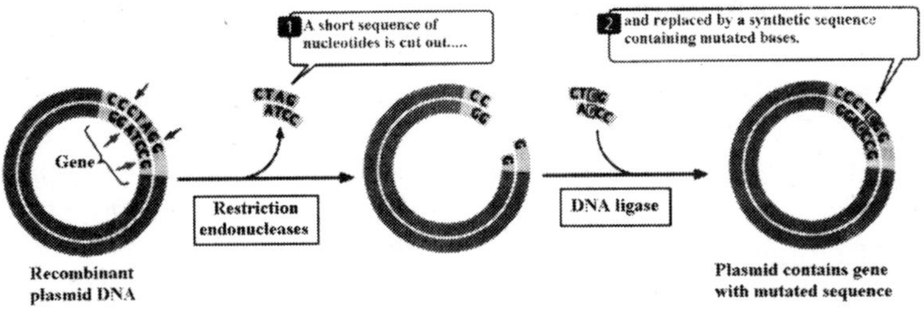

Figure 2.36 In site-directed mutugenesis. *Restriction enzymes cut out a short sequence of nucleotides that is then replaced by a synthetic mutated DNA sequence.*

If appropriate restriction sites are not available, **oligonulceotide-directed mutagenesis** is used (**Figure 2.37**). In this method, a single-stranded oligonulceotide is produced that differs from the target sequence by one or a few bases. Because they differ by only a few bases, the target DNA and the oligonulceotide will pair under the appropriate conditions. When successfully paired with the target DNA, the oligonucleotide can act as a primer to initiate DNA synthesis, which produces a double-stranded molecule with a mismatch in the primer region.

When this DNA is transferred to bacterial cells, bacterial enzymes will repair the mismatched bases. About half the time the normal bases will be changed into mutant bases. The bacteria are then screened for the presence of the mutant gene.

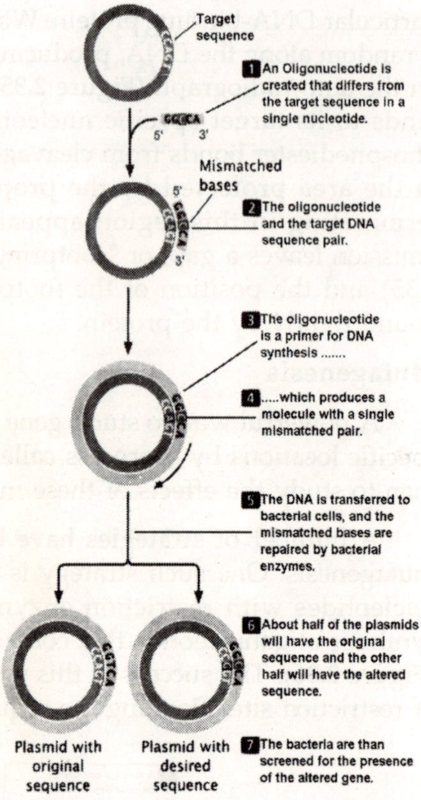

Figure 2.37 Oligonucleotide –directed mutagenesis. *Mutation can be created at any specific site in a gene whose wild -type sequence is already known. The known sequence is used to chemically synthesize short DNA segments called oligonucleotides.*

Transgenic animals

The oocytes of mice and other mammals are large enough for DNA to be injected into them directly. Immediately after penetration by sperm, a fertilized mouse egg contains two pronuclei, one from the sperm and other from the egg; these pronuclei later fuse to from the nucleus of the embryo. Mechanical devices are used to manipulate extremely fine, hollow glass needles to inject DNA directly into one of the Pronuclei of a fertilized egg (**Figure 2.38**). Typically, a few hundred copies of cloned, linear DNA are injected into a pronucleus, and, in a few of the injected eggs, copies of the cloned DNA integrate randomly into one of the chromosomes through a process called nonhomologous recombination. After injection, the embryos are implanted in a pseudo pregnant female, a surrogate mother that has been physiologically prepared for pregnancy by mating with a vasectomized male. Only about 10% to 30% of the eggs survive and, of those that do survive, only few may have a copy of the cloned DNA stably integrated into a chromosome.

Of several hundred embryos which are injected and implanted, there is a good chance that one or more mouse whose chromosomes contain the foreign DNA will be born. Moreover, because the DNA

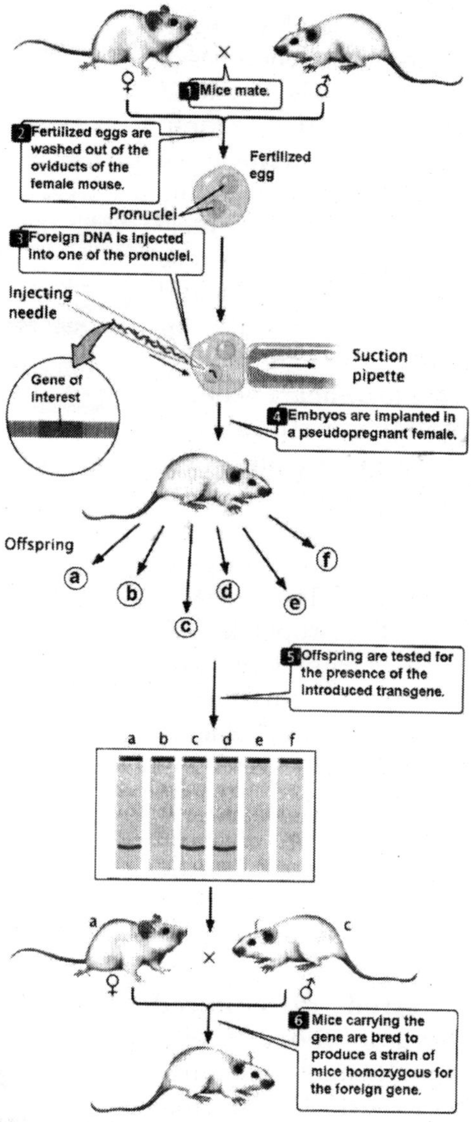

Figure 2.38 Construction of transgenic animals. *Animals develop altered genetic structure.*

was injected at one-cell stage of the embryo, these mice usually carry the cloned DNA in every cell of their body, including their reproductive cells and will therefore pass the foreign DNA to their progeny. Through interbreeding, a strain of mice that is homozygous for the foreign gene can be created. Animals that have been permanently altered in this way are said to be *transgenic* and the foreign DNA that they carry is called a **transgene**.

Transgenic mice have proved useful in the study of gene function. Injecting a copy of the SRY gene into XX embryos and observing that these mice developed as males obtained proof that the SRY gene is the male-determining gene in mice. In addition, a number of transgenic mouse strains that serve as experimental models for human genetic diseases have been created by injecting mutated copies of genes into mouse embryos.

Knockout mice

A particularly useful variant of the transgenic approach is to produce mice in which a normal gene has been disabled. The phenotypes of these animals, called **knockout mice**, help geneticists to determine the function of a gene. The creation of knockout mice begins when a normal gene is cloned in bacteria and then "knocked out", or disabled. There are number of ways to disable a gene but a common method is to insert a gene called neo, which confers resistance to the antibiotic G418, into the middle of the target gene (**Figure 2.39**). The insertion of neo both disrupts (knocks out) the target gene and provides a convenient marker for finding copies of the disabled gene. In addition, a second gene, usually the herpes simplex viral thymidine kinase (tk) gene, is cloned adjacent to the disrupted gene. The disabled gene is then transferred to cultured embryonic mouse cells, where it may exchange places with the normal chromosomal copy through homologous recombination.

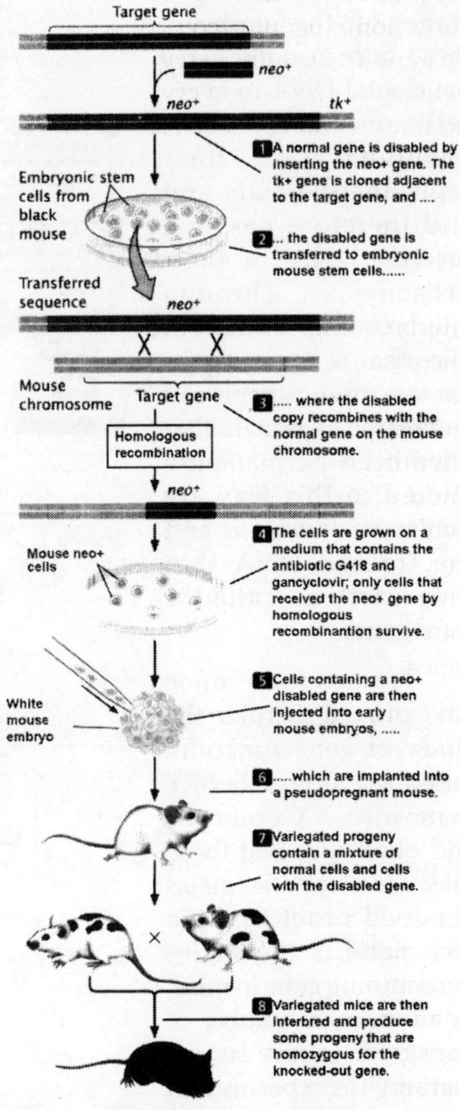

Figure 2.39 Knockout mice possess a genome in which a gene has been disabled.

After the disabled gene has been transferred to the embryonic cells, the antibiotic G418 is used to screens the cells. Only cells with the disabled gene containing the *neo* insert will survive. The frequency of non-homologous recombination is higher than that of homologous recombination and the disabled copy replaces the intact target gene only through homologous recombination. A method to select for the rarer homologous recombinants is required. The presence of the viral tk gene makes the cells sensitive to gancyclovir. Thus, transfected cells that grow on medium containing G418 and gancyclovir will contain the neo gene (disabled target gene) but not the adjacent tk gene and contain the desired homologous recombinants. The nonhomologous recombinants (random insertions) will contain both the neo and the tk genes and these transfected cells will die on the selection medium owing to the presence of gancyclovir. The surviving cells are injected into an early-stage mouse embryo, which is then implanted into a pseudo pregnant mouse. Cells in the embryo carrying the disabled gene and normal embryonic cells carrying the wild type gene will develop together, producing a chimera- a mouse that is a genetic mixture of the two cell types. The chimeric mice can be identified easily if the injected embryonic cells came from a black mouse and the embryos into which they are injected come from a white mouse; the resulting chimeras have a variegated black and white fur. The chimeras can then be interbred to produce some progeny that are homozygous for the knockout gene. The phenotypic effects of disabling a particular gene can be observed in these homozygous mice.

Although these are recent innovation, knockout mice have become important subjects for research in a number of fields. They have been used to study genes implicated in immune function, development, aetiology, and human behaviour.

Applications of Recombinant DNA Technology

In addition to providing valuable new information about the nature and function of genes, recombinant DNA technology has many practical applications. These applications include the production of pharmaceuticals and other chemicals, specialized bacteria, agriculturally important plants and genetically engineered farm animals. The technology is also used extensively in medical testing and, in a few cases, is even being used to correct human genetic defects. Hundreds of firms now specialize in developing products through genetic engineering, and many large multinational corporations have invested enormous sum of money in recombinant DNA research. Recombinant DNA technology is also frequently used in criminal investigations and for the identification of human remains. For example, the remains of people who died in the collapse of the World

Trade Centre on September 11, 2001, were identified through DNA comparisons using recombinant DNA technology.

Pharmaceuticals

The first commercial products to be developed using recombinant DNA technology were pharmaceuticals used in the treatment of human disease. In 1979, the Eli Lilly Corporation marketed human insulin used in the treatment of diabetics by recombinant DNA technology to replace insulin isolated from the pancreas of farm animals slaughtered for meat. Although animal insulin worked well for many diabetics, being a foreign protein many people suffered allergic reactions. Pharmaceuticals produced through recombinant DNA technology include human growth hormone (for children with growth deficiencies), clotting factors (for hemophiliacs), and tissue plasminogen activator (used to dissolve blood clots in heart attack patients).

Specialized Bacteria

Bacteria play an important role in many industrial processes, including the production of ethanol from plant material, the leaching of minerals from ore, and the treatment of sewage and other wastes. The bacteria engaged in these processes are being modified by genetic engineering so that they work more efficiently. New strains of technologically useful bacteria are being developed that will break down toxic chemicals and pollutants, enhance oil recovery, increase nitrogen uptake by plants, and inhibit the growth of pathogenic bacteria and fungi.

Agricultural products

Recombinant DNA technology has had a major effect on agriculture, where it is now used to create crop plants and domestic animals with valuable traits. For many years, plant pathologists had recognized that plants infected with mild strains of viruses are resistant to infection by virulent strains. Using this knowledge, geneticists have created viral resistance in plants by transferring genes for viral proteins to the plant cells. A genetically engineered squash, called freedom II, carries genes from the watermelon mosaic virus 2 and the zucchini yellow mosaic virus that protect the squash against viral infections.

Another objective has been to genetically engineer pest resistance into plants to reduce dependence on chemical pesticides.

A protein toxin from the bacterium *Bacillus thuringiensis* selectively kills the larvae of certain insect pests but is harmless to wildlife, humans and many other insects. The toxin gene has been isolated from the bacteria, linked to active promoters and transferred into corn, tomato, potato and cotton plants. The gene produces the insecticidal toxin in the plants, killing caterpillars that feed on the plant.

Recombinant DNA technology has also permitted the development of herbicide resistance in plants. A major problem in agriculture is the control of weeds, which compete with crop plants for water, sunlight, and nutrients. Although herbicides are effective at killing weeds, they can also damage crop plants. Genes that provide resistance to broad-spectrum herbicides have been transferred into tomato, soyabean, cotton, oilseed rape and other commercially important crops. When the fields containing these crops are sprayed with herbicides, the weeds are killed but the genetically engineered plants are unaffected. In 1999, more than 21 million hectares (1hectare=2.471 acres) of genetically engineered soyabeans and 11 million hectares of genetically engineered corn were grown throughout the world.

Recombinant DNA techniques are also being applied to domestic animals. For example, recombinant bovine growth hormone has been used to increase milk production of cattle. Transgenic animals are being developed to carry genes that encode pharmaceutical products. A gene for human clotting factor VIII, which is used to treat haemophiliacs, has been linked to the regulatory region of the sheep gene for β-lacto globulin, a milk protein. The fused gene was injected in sheep embryos, creating transgenic sheep that produce the human clotting factor in their milk. A similar procedure was used to transfer a gene for a1- antitrypsin, generating $100,000 worth of a1- antitrypsin per year per sheep.

The genetic engineering of agricultural products is controversial. One area of concern focuses on the potential effects of releasing novel organisms produced by genetic engineering into the environment. There are many examples in which non-native organisms released into a new environment have caused ecological disruption because they are free of predators and other natural control mechanisms. Genetic engineering normally transfers only small sequences of DNA, relative to the large genetic differences that often exist between species but even small genetic differences may alter ecologically important traits that might affect the ecosystem.

Another area of concern is that transgenic organisms may hybridize with native organisms and transfer their genetically engineered traits. For example, herbicide resistance engineered into crop plants might be transferred to weeds, which would then be resistant to the herbicides that are now used for their control. The results of some studies have demonstrated gene transfer between engineered plants and native plants but the extent and effect of this transfer are uncertain. Other concerns focus on health-safety issues associated with the presence of engineered products in natural foods. Labelling of all genetically engineered foods that contain transgenic DNA or protein is mandatory in many countries.

On the other hand, the use of genetically engineered crops and domestic animals has potential benefits. Genetically engineered crops that are pest resistant have the potential to reduce the use of environmentally harmful chemicals and research findings indicate the lower amounts of pesticides are used in the United States as a result of the adoption of transgenic plants. Transgenic crops also increase yields, providing more food per acre, which reduces the amount of land that must be used for agriculture.

Oligonulceotide Drugs

A recent application of DNA technology has been the development of oligonulceotides as drugs. These are short sequences of synthetic DNA or RNA molecules that can be used to treat diseases (**Figure 2.40**). Antisense oligonulceotides are complementary to undesirable RNAs, such as viral RNA. When added to a cell, these antisense DNAs bind to the viral mRNA and inhibit its translation.

Single-stranded DNA oligonucleotides bind tightly to other DNA sequences, forming triplex DNA molecules (**Figure2.40**). The

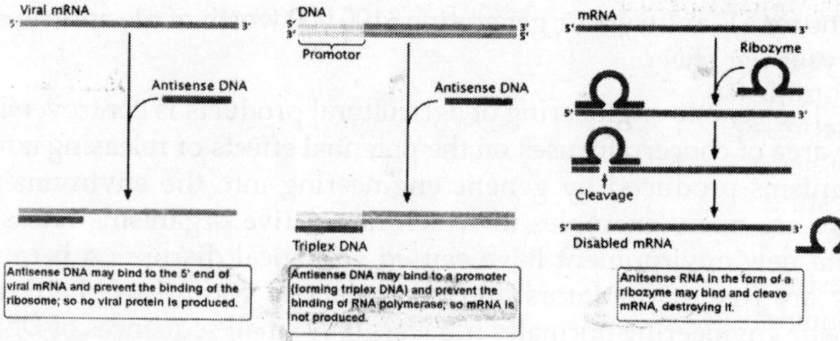

Figure 2.40 Oligonucleotide drugs are short sequences of DNA or RNA that can be used to treat diseases.

formation of triplex DNA interferes with the binding of RNA polymerase and other proteins required for transcription. Other oligonucleotides are ribozymes and RNA molecules that function as enzymes. These compounds bind to specific mRNA molecules and cleave them into fragments, destroying their ability to encode proteins. Several oligonulceotide drugs are already being tested for the treatment of AIDS and cancer.

Genetic Testing

The identification and cloning of many important disease causing human genes has allowed the development of probes for detecting disease-causing mutations. Prenatal testing is already available for several hundred genetic disorders. Additionally, presymptomatic genetic tests for adults and children are available for an increasing number of disorders.

The growing availability of genetic tests raises a number of ethical and social issues. For example, is it ethical to test for genetic diseases for which there is no cure or treatment? Huntington's disease, an autosomal dominant disorder that appears in middle age, causes slow physical and mental deterioration and eventually death. No effective treatment is currently available. If one parent is affected, a child has a 50% chance of inheriting the gene for Huntington's disease and eventually getting the disorder. Tests are now available that make it possible to detect persons who carry the defective gene but is it beneficial to tell a young person that he or she has the defective gene and will get the disease later in life?

Where a negative test might provide peace of mind, a person testing positive will suffer from despair and depression. Many people at risk for Huntington's disease want predictive testing, saying that the uncertainty of not knowing is more debilitating than to have certain knowledge that they will get it and, in fact, a number of medical centres now offer predictive testing. However, some people have committed suicide, and others have had to be hospitalized for depression after learning that they have the defective gene. Results of several studies indicate that most people who undergo predictive testing for Huntington's disease are able to cope with the information.

Other ethical and legal questions concern the confidentiality of test results. Who should have access to the results of genetic testing? Should insurance companies be allowed to use results from such tests to deny coverage to healthy people who are at risk for genetic diseases? Should relatives who also might be at risk be informed of the results of genetic testing?

Other concerns focus on whether the cost of genetic testing justifies the benefits. In some diseases, genetic tests provide clear benefits because early identification allows for better treatment. In the case of phenylketonuria (an autosomal recessive disorder that can cause mental retardation) its identification in infants, can prevent mental retardation by the administration of special diet. Because of this obvious benefit and the low cost of testing for this disorder, all newborns in United States and many other countries require are tested for PKU.

Predictive testing for colorectal cancer and breast cancer also may be beneficial for people at risk, because finding these cancers early improves the chances for successful treatment. Patients with genes that predispose to cancer may require more aggressive treatment than do patients with spontaneously arising cancers.

Another set of concerns is related to the accuracy of genetic tests. For many genetic diseases, the only predictive tests available are those that identify a predisposing mutation in DNA but dozens or hundreds of different mutations may cause many genetic diseases. Probes that detect common mutations can be developed but these do not detect rare mutations and hence the tests give false negative results. The only definite way to identify all predisposed persons is sequencing the entire gene. These questions and concerns are currently the focus of intense debate by ethicists, physicians, scientists, and patients.

Gene therapy

Perhaps the ultimate application of recombinant DNA technology is gene therapy, the direct transfer of genes to treat disease. When the first recombinant DNA experiments with bacteria were announced, many researchers recognized the potential for using this new technology for the treatment of patients with genetic diseases. But, before recombinant DNA could be used on humans, a number of difficult obstacles had to be overcome. The genes responsible for particular genetic diseases needed to be located and cloned and special vectors had to be developed that would reliably and efficiently deliver genes to human cells.

The first effort in gene therapy was made in 1990 by W. French Anderson and his colleagues at the U.S. National Institutes of Health (NIH). They transferred a functional gene for adenosine deaminase to a young girl with severe combined immunodeficiency disease, an autosomal recessive condition that produces impaired immune function.

The success of gene therapy depends on ability of introduced gene to produce a therapeutic protein. A number of different methods for transferring genes into human cells are currently under development. Commonly used vectors include genetically modified retroviruses, adenoviruses, and adeno-associated viruses (**Table 2.9**). Genes are transferred by adding viruses containing recombinant genes to the patient's cells (such as white cells), and then reintroducing these cells back into the patient. In other cases, vectors are injected directly into the patient.

Table 2.9: Vectors used in gene therapy

Vector	Advantages	Disadvantages
Retrovirus	Efficient transfer	Transfers DNA only to dividing cells, inserts randomly; risk of producing wild-type viruses
Adenovirus	Transfers to nondividing cells	Causes immune reaction
Adeno-associated virus	Does not cause immune reaction	Holds small amount of DNA; hard to produce
Herpes virus	Can insert into cells of nervous system; does not cause immune reaction	Hard to produce in large quantities
Lentivirus	Can accommodate large genes	Safety concerns
Liposome's and other lipid-coated vectors	No replication; does not stimulate immune reaction	Low efficiency
Direct injection	No replication; directed toward specific tissues	Low efficiency; does not work well within some tissues
Pressure treatment	Safe, because tissues are treated outside the body and then transplanted into the patient	Most efficient with small DNA molecules
Gene gun (DNA coated on small gold particles and shot into tissue)	No vector required	Low efficiency

In spite of the growing number of clinical trials for gene therapy, significant problems remain in transferring foreign genes into human cells, getting them expressed, and restricting immune responses to the gene products and the vectors used to transfer the genes. There are also heightened concerns about safety. In 1999 a patient participating in a gene-therapy trial had a fatal immune reaction after he was injected with a viral vector carrying a gene to treat his metabolic disorder.

Up till now gene therapy has targeted only non-reproductive, somatic cells in which it provides positive benefits to patients without affecting the genes of future generations. Gene therapy that alters reproductive or germ-line cells (termed germ-line gene therapy) is technically possible but raises a number of ethical issues because it has the capacity to alter the gene pool of future generations.

Gene Mapping

A significant contribution of recombinant DNA technology has been to provide numerous genetic markers that can be used in gene mapping. One group of markers used in gene mapping comprises restriction fragment length polymorphisms (RFLPs). RFLPs are variations (polymorphisms) in the patterns of fragments produced when DNA molecules are cut with the same restriction enzyme. If DNA from two persons is cut with the same restriction enzyme and different patterns of fragments are produced (**Figure 2.41**), these persons must

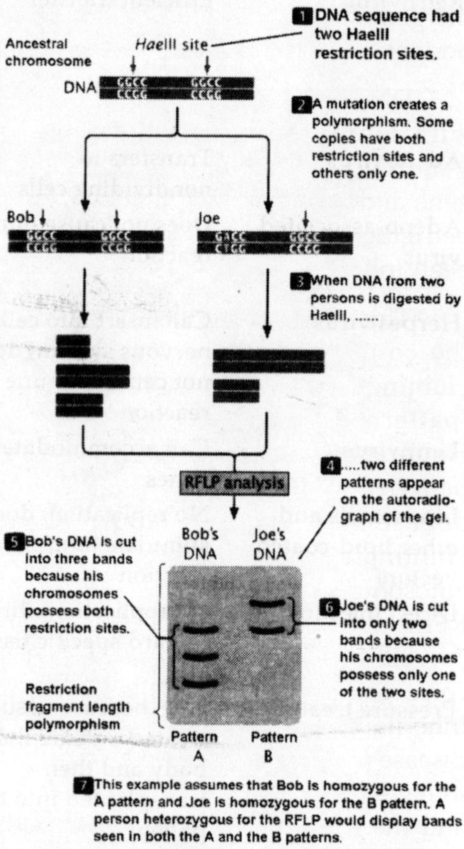

Figure 2.41 Restriction fragment length polymorphisms are genetic markers that can be used in mapping.

possess differences in their DNA sequences. These differences are inherited and can be used in mapping, similar to the way in which allelic differences are used to map conventional genes.

Traditionally, gene mapping has relied on the use of genetic differences that produce easily observable phenotypic differences. Unfortunately, because multiple genes and the environment influence most traits, the number of traits with a simple genetic basis suitable for use in mapping is limited. RFLPs provide a large number of genetic markers that can be used for mapping. Until recently the chromosomal location of the gene for Huntington disease, an autosomal dominant disease was unknown. James Gusella and his team set out to determine the location of the Huntington's disease gene, in the hope that, when the gene was found, its biochemical basis could be determined and possible treatments could be suggested.

DNA was collected from members of the largest known family with Huntington's disease, who live near Lake Maracaibo in Venezuela. They searched for co-inheritance of the disease-causing gene and an RFLP with a known chromosomal location in this large pedigree. The rationale was that the disease gene and RFLP have been inherited together and must be physically linked.

This approach is summarized in **Figure 2.41**, which illustrates the co-inheritance to two traits: (1) the presence or absence of Huntington's disease and (2) the type of restriction pattern produced (pattern A or C). In this family, the father is heterozygous for Huntington's disease (Hh) and is also heterozygous for a restriction pattern (AC). From the father, each child inherits either a Huntington's disease allele (H) or a normal allele (h); any child inheriting the Huntington's disease allele develops the disease, because it is an autosomal dominant disorder. The child also inherits one of the two RFLP alleles from the father, either A or C, which produces the corresponding RFLP pattern. In **Figure 2.42a**, there is no correspondence between the inheritance of the RFLP pattern and the inheritance of the disease: children who have inherited Huntington's disease (and therefore the H allele) from their father are equally likely to have inherited the A or C RFLP pattern. In this case, the H allele and the RFLP alleles segregate randomly, so we know that they are not closely linked.

Figure 2.42b, on the other hand, shows that every child who inherits the C pattern from the father also inherits Huntington's

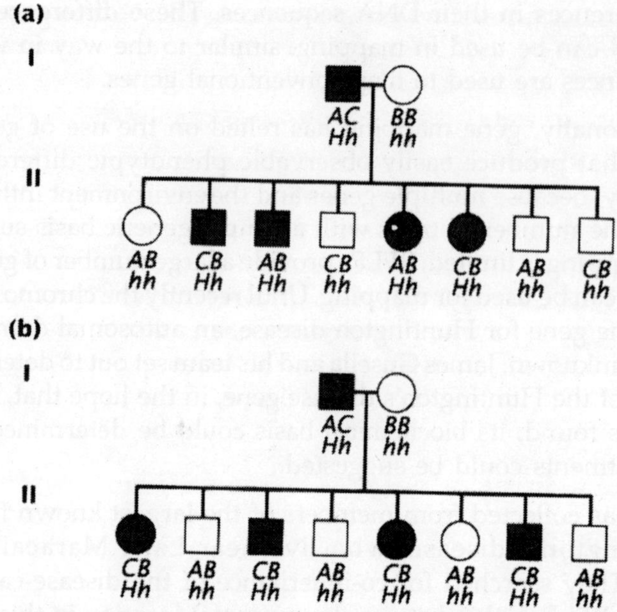

Figure 2.42 Restriction fragment length polymorphisms can be used to detect linkage. *In this hypothetical pedigree, the father and half of the children are affected (dark circles and squares) with Huntington's diseases, an autosomal dominant disease. The father is heterozygous (Hh) and will pass the chromosome with the Huntington's gene to approximately half of his offsprings. The father is also heterozygous for RFLP alleles A and C; each child receives one of these two alleles from the father. The mother is homozygous for RFLP allele B, so all children receive the B allele from her. (a) In this case, there is no correspondence between the inheritance of the RFLP allele and inheritance of the disease-children with the disease are just as likely to carry the A allele as they are the C allele. Thus the disease gene and RFLP alleles segregate independently and are not closely linked. (b) In this case there is a close correspondence between the inheritance of the RFLP alleles and the presence of the disease- every child who inherits the C allele from the father also has the disease. This correspondence indicates that the RFLP is closely linked to the Huntington gene.*

disease (and therefore the H allele), because the locus for the RFLP is closely linked to the locus for the disease-causing gene. The chromosomal location of the RFLP provides a general indication of the disease-causing locus. An examination of the co-segregation of other RFLPs from the same region can precisely determine the location of the gene. Actual RFLP patterns and part of the Huntington disease gene are shown in **Figure 2.43**.

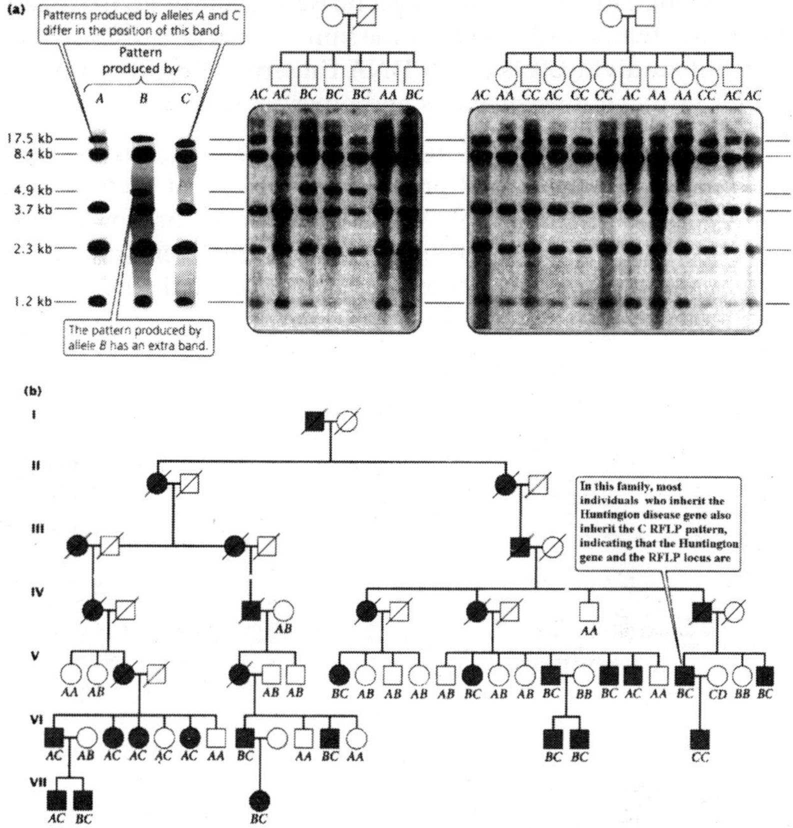

Figure 2.43 ***Restriction fragment length polymorphisms were used to map the Huntington-disease gene to chromosome 8.*** *(a) Autoradiograph showing different banding patterns revealed by cutting the DNA with HindIII and using a probe for chromosome 8. The RFLP an allele produces five bands. The C allele also produces five bands, but the first band is just below the first band produced by the A allele; AC heterozygotes have both bands, which are very close together. The B allele has an extra band representing a 4.9-kb fragment. (b) Partial pedigree of large family from Lake Maracaibo. Dark symbols represent family members with Huntington's disease; the RFLP genotypes are indicated below each person represented in the pedigree. Notice that persons with the disease carry the C allele, indicating that the sequences on chromosome 8 revealed by the probe are closely linked to the Huntington-disease gene.*

DNA Fingerprinting

Restriction fragment length polymorphisms are often found in non-coding regions of DNA and are therefore frequently quite variable in humans. Two randomly chosen people will differ at many RFLPs and, if enough RFLPs are examined, no two people (with the exception of identical twins) will be exactly the same. The use of DNA sequences to identify individuals is called **DNA fingerprinting**. It is a powerful tool for criminal investigations and other forensic applications.

In a typical application, DNA fingerprinting can be used to confirm that a suspect was present at the scene of a crime (**Figure 2.44**) by using sample of DNA from blood, semen, hair or other body tissues collected from the scene. If the sample is very small, PCR is used to amplify DNA, which is then compared to the DNA samples collected from one or more suspects.

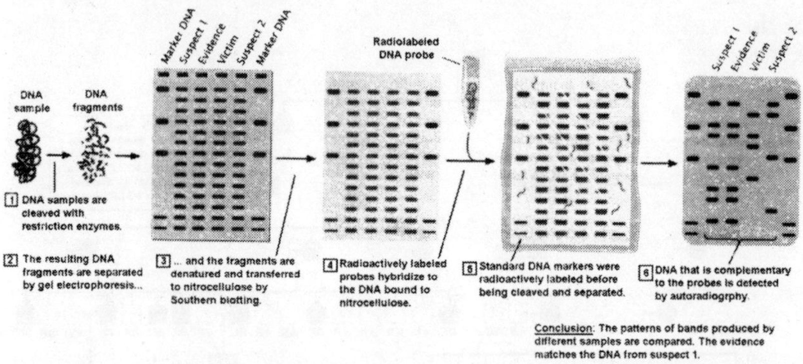

Figure 2.44 *The use of DNA sequences to identify a person is called DNA fingerprinting.*

DNA samples are cut with one or more restriction enzymes and the resulting DNA fragments are separated by gel electrophoresis. The fragments in the gel are denatured, transferred to nitrocellulose by Southern blotting, hybridized and detected by autoradiography. There after DNA banding pattern from the sample collected at the crime scene is compared with the patterns produced by DNA from the suspects.

The probes used in DNA fingerprinting detect highly variable regions of the genome; so the chances of DNA from two people producing exactly the same banding pattern are low. When several probes are used in the analysis, the probability that two people have the same set of patterns becomes small (unless they are identical twins).

The probes most commonly used in DNA fingerprinting are complementary to short sequences repeated in tandem that are widely found in the human genome. People vary greatly in the number of copies of these repeats; thus these polymorphisms are termed **variable number of tandem repeats** (VNTRs).

Since its introduction in 1980s, DNA fingerprinting has helped convict a number of suspects in murder and rape cases. Suspects in other cases have been proved innocent when their DNA failed to match that from the crime scenes. Initially there was some controversy over calculating the odds of a match (the probability that two people could have the same pattern) and concerns about quality control (such as the accidental contamination of samples and the reproducibility of results) in laboratories where DNA analysis is done. In spite of the controversy, DNA fingerprinting has become an important tool in forensic investigations.

DNA fingerprinting has also been used to provide information about the relationships and sources of other organisms. For example, DNA fingerprinting was used to determine that several samples of anthrax mailed to different people in 2001 were all from the same source.

CHAPTER

3 Major Histocompatibility Complex

The immune response requires initial processing and presentation of antigen to T- cells by **antigen presenting cells (APC's)**. Antigens get broken down into peptides, which are then displayed in association with **major histocompatibility complex (MHC)** components; the MHC antigen complex is recognised by T - cell receptor (TCR) on T cells. MHC is a generic name for the group of genes involved in tissue recognition of self and non-self. In humans it is termed **Human Leukocyte Antigens** or HLA. The nomenclature for the MHC in different species is shown in **Table 3.1**.

Table 3.1: Nomenclature of MHC in different species

Species	MHC Designation	
Mouse	H -2	Histocompatibility System - 2
Human	HLA	Human Leukocyte Antigens
Rabbit	RLA	Rabbit Leukocyte Antigens
Guinea Pig	GPLA	Guinea Pig Leukocyte Antigens
Chimpanzee	ChLA	Chimpanzee Leukocyte Antigen
Cattle	BoLA	Bovine Leukocyte Antigens

MHC research started in the beginning of the 20th century when Little and Tyzzer analysed the fate of tumours transplanted between different mouse strains. In 1916, they demonstrated that several dominant genes influence the outcome of allogenic tumour grafts. Ten years later Bover showed that skin graft between identical twins were not rejected, further demonstrating that genetic factors determine transplant outcome. The search for these genetic factors continued till Gorer identified MHC in the mouse in 1944. Snell while studying MHC genes in transplantations between congenic mice proposed the name, histocompatibility genes (H). The genes were designated antigen II by Gorer and the blended name became H-2, which is even used today for the mouse MHC.

The human MHC was first described in the 1950s with the recognition of antibodies in the sera of multiply transfused patients and multiparous women, which reacted with leukocytes of other members of the population. The antigens involved were termed Human Leukocyte Antigens (HLA). It was quickly realised that they play a central role in the success or failure of organ transplantation, and this provided impetus to study these antigens in greater detail. Interest in the MHC increased further as its role in the regulation of immune responses, and its association with diseases became apparent in the 1970s. Initially, the serological techniques identified only two polymorphic gene loci, the HLA-A and HLA-B series, but soon other loci were identified. In the 80s introduction of molecular techniques revolutionised HLA typing and presently over 500 different alleles of MHC genes have been identified, far more than for any other genetic region. The extensive polymorphism of these loci makes it very unlikely, for any two individuals in a population to have the same combination of MHC alleles, which is why there is rejection of transplanted organs from unrelated donors.

The genes for the HLA antigens are located on the short arm of chromosome 6 in the band 6p 21.3. It takes up at least 4 million nucleotide pairs. It is arbitrarily subdivided into class I, II and class III gene clusters. These gene clusters constitutes more than 120 loci Class I and II gene products are structurally similar and are cell surface glycoproteins involved in antigen presentation to T cells. Class III gene code for heterogeneous proteins including several components of the complement system. These are structurally unrelated to class I and II gene products and are grouped with them only because they are located in the same chromosomal region. Class I antigens are expressed on all somatic nucleated cells, whereas class II is expressed on restricted population of cells, which are usually capable of antigen presentation. The complement system and other soluble molecules constitute the class III component of the MHC molecule.

MHC Class I Molecule

MHC class I molecules consist of MHC encoded heavy chain and β2 micro-globulin, the gene for which is located on chromosome 15 in humans. Amongst the class I loci, HLA-B and –C loci are more centromeric than HLA- A locus and span about 85 Kb in length. HLA –A locus is more telomeric in position and forms a cluster with class I related genes (**Figure 3.1**).

Of 20-odd genes found in the HLA class I region, only 6 are expressed at the protein level. These are further classified as classical and non-classical genes according to their tissue specific expression, level of polymorphism and function. The classical class I genes are HLA -A, - B, -C while the non-classical genes are HLA-E, -F and -G. The non-expressed loci are either full-length pseudo genes or gene fragments.

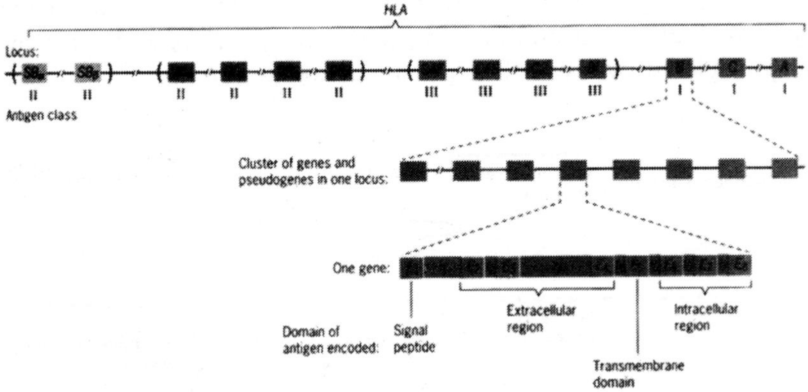

Figure 3.1 Organization of the major histocompatibility complex (HLA) on human chromosome 6. *The relative positions of mapped loci within this huge gene complex are shown at the top. The class of histocompatibility antigen encoded by genes in each locus is indicated just below the map. Each locus within the HLA complex contains several genes and pseudogenes (center). The structure of a typical class I gene is shown at the bottom. Note that different exons encode different functional domains of the polypeptide gene product, key E= exon; I = intron.*

Structure of Class I Molecule

Class I molecules are integral membrane glycoproteins made of a heavy or α chain of about 44 kilo Daltons (kD) which is non-covalently associated with β 2 microglobulin, a 12 kD polypeptide (**Figure 3.2**). β2 micro globulin is also found as a soluble protein in serum. The heavy chain consists of 3 extra cellular domains, α1, α2 and α3, a transmembrane region and a cytoplasmic domain. The α3 domain is closest to the plasma membrane. β2 microglobulin is associated with the extra cellular portion of the α chain. The three extra cellular domains are composed of 90 amino acids each. α2 and α3 domains have inter chain disulphide bonds enclosing loops of 63 and 86 amino acids respectively and the α3 domain shares features with immunoglobulin (Ig) constant region. A glycosylated asparagine residue is present at position 86 of α3 domain. The transmembrane region consists of 23 hydrophobic uncharged amino acid residues, which probably assume an α - helical conformation. A cluster of about 5 basic amino acids (arginine and lysine) is located immediately C terminal to the membrane and is typical of membrane-bound proteins. They are thought to anchor the polypeptide chain in the membrane by interacting with negatively charged phospholipid head groups of the inner membrane.

The hydrophobic cytoplasmic domain consists of 30 amino acids; of which 50 are polar amino acids, particularly serine, and some are phosphorylated by a cAMP dependent protein kinase.

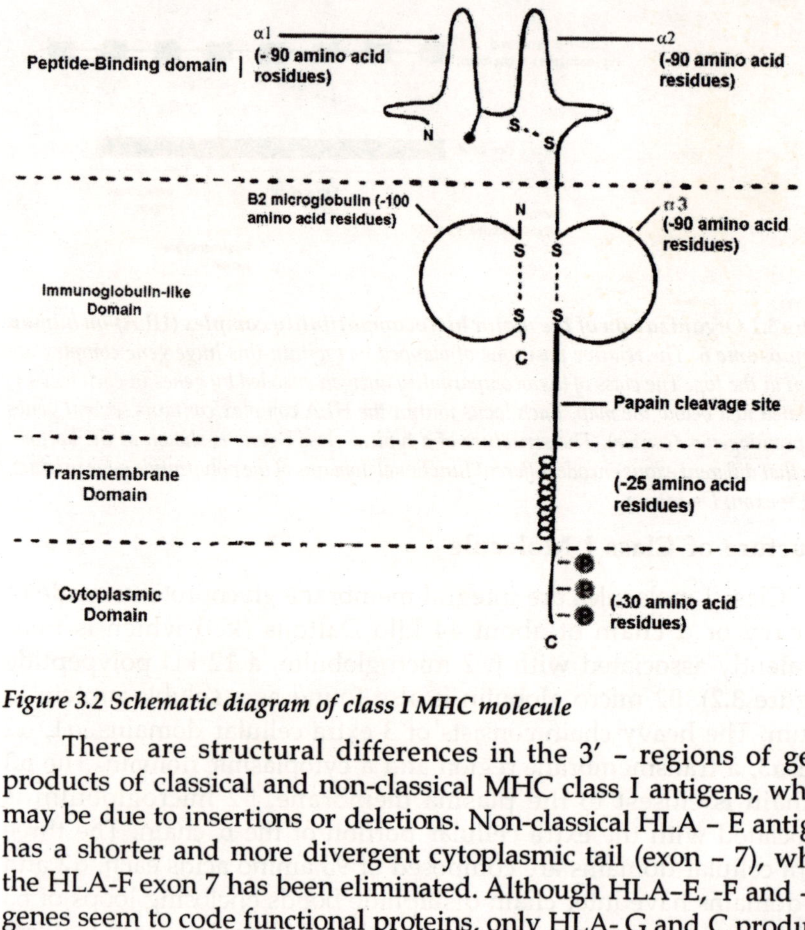

Figure 3.2 Schematic diagram of class I MHC molecule

There are structural differences in the 3′ – regions of gene products of classical and non-classical MHC class I antigens, which may be due to insertions or deletions. Non-classical HLA – E antigen has a shorter and more divergent cytoplasmic tail (exon – 7), while the HLA-F exon 7 has been eliminated. Although HLA–E, -F and – G genes seem to code functional proteins, only HLA- G and C products are detected in the normal cell population at low but significant levels. The non-classical class I genes are fully expressed when they are transfected into a lymphoblastoid cell lines lacking HLA- A, -B and -C but expressing β2 microglobulin. Non-classical gene products are less polymorphic than the classical A and B loci, although alternate splicing has been discovered for the HLA-G gene resulting in five different isoforms i.e. G1, G2, G3, G4, G5.

The classical class I HLA genes are expressed on most nucleated cells with the exception of cells of the central nervous system, especially neurons and astrocytes, skeletal and smooth muscle cells, parathyroid cells, exocrine pancreatic cells and corneal epithelium. None of the trophoblast cell subpopulations, which constitute the materno-fetal interface during pregnancy, express these molecules. Both male and female germinal cells are also devoid of classical class I antigens. The placenta and extra villous membranes express non-classical class I antigens.

X - Ray Crystal Structure

The first X-ray crystal structure of a MHC class I molecule was reported in 1987 for human HLA A2. The molecule was purified from a human lymphoblastoid B cell line after papain digestion of the plasma membranes. The structure revealed that α 1 and α 2 domains together form one structural domain (α1 / α 2) supported by the α3 domain and β2m. The α1 / α2 domain consists of two nearly anti-parallel α-helices above a floor of eight β-pleated strands. The antigen-binding site lies in a groove in the most membrane distal portion of the molecule, between the α -helices (**Figure 3.3**).

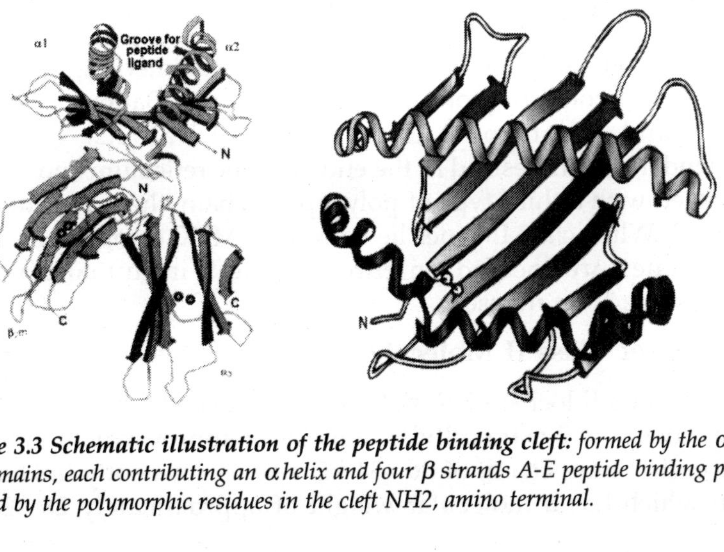

Figure 3.3 Schematic illustration of the peptide binding cleft: *formed by the α1 and α2 domains, each contributing an α helix and four β strands A-E peptide binding pockets formed by the polymorphic residues in the cleft NH2, amino terminal.*

Regulation of MHC Class - I Gene Transcription

Although MHC class I gene products are constitutively expressed on all nucleated cells, the level of expression varies widely and is highly regulated. In majority of instances, regulation is at the transcriptional level. For both class I and class II genes *cis* acting DNA elements (promoters & enhancers) and transacting factors that bind to these DNA elements, have been identified.

The *cis* regulatory elements are found 5′ to the transcription initiation sites. The class I regulatory complex (CRC) has several conserved sets of sequences like Enhancer A which has binding sites for the transcription factor NF-kB and interferon stimulated response element or ISRE, which has been shown in other genes to bring about interferon mediated regulation of gene expression. Various trans acting proteins binding to these sequences have been identified; however, much remains to be understood about their role in the expression of the HLA class I genes.

A number of cytokines such as IFN - α , -β, and -γ, and TNF-α have been shown to affect class I and II expression. IFN - α induces specific nuclear binding proteins, which bind to regulatory sequences and up regulate transcription of class - I MHC gene. Interferons also increase MHC class II expression. However, regulation by other lymphokines is limited. MHC expression has also been shown to be down regulated by certain modulators like corticosteroids and prostaglandins.

MHC Class II Molecule

The products of class II genes are heterodimeric glycosylated polypeptides consisting of an α chain of 34,000 Daltons molecular weight and a light or β chain of 28,000 Daltons. These are independently synthesized in the endoplasmic reticulum and become associated with a third type of polypeptide chain, the invariant chain (Ii chain). While class II genes lie within the MHC complex, invariant chain genes are located on chromosome 5 in humans and on chromosome 18 in mouse.

Structure of Class II Molecule

The class II loci are DR, DP, DQ, DZ, DO and DW. The α and β chains consist of 2 extra cellular domains of approximately 90 amino acids each. The α chain, of 30 -34 kD size, is covalently linked to β chain which has a molecular weight of approximately 28 -29 kD

(**Figure 3.4**). Molecular mapping of class II MHC genes has revealed additional chains identified both in mouse and humans. In the human DR region there is more than one functional β chain gene; the number of genes varies with different class II haplotypes. All these products can associate with the α chain to form class II cell surface molecules, thus providing a mechanism for increasing the polymorphism of the DR molecule. Pseudogenes of unknown function also exist. Genes which encode molecules that function as **peptide transporters** are also mapped within class II region. These proteins transport peptides from cytoplasm to endoplasmic reticulum, enabling the peptide to interact with newly-synthesised class I molecules.

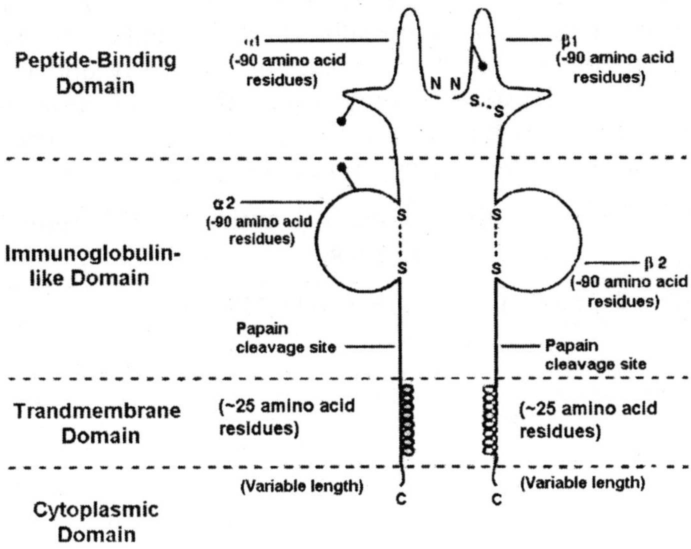

Figure 3.4 Schematic diagram of a class II MHC molecule

The membrane distal domains of class II molecule, the α 1 and β1 domain, form the antigen binding cleft for processed antigens. The class II genes are highly polymorphic with, most of the polymorphism occurring in DRβ and DQβ chains; DPβ chains are less polymorphic. Of the α chains, only DQ α is polymorphic, DRα being invariant in all the alleles.

All class II genes are prefixed by a letter for the sub region (N, O, P, Q, R) and then the letters A or B for α or β chains; an Arabic number for the different genes follows this. For example, DQB1 designates the gene that encodes the expressed β chain of the DQ molecule and DQA1 codes for α chain of DQ region. Alleles for each gene are numbered using four digits. The first two digits describe the most closely associated serologic specificity and the next two the subtypes within a serological specificity. An asterisk is used to separate genes and alleles. For example DR4 subtypes would be written as DRB1 * 0401, DRB * 0403 and DRB1 *0405, while DQB1 * 0302 is a DQB subtype of the DQ3 specificity.

Distribution of Class II molecule

Compared to MHC class I molecules, tissue distribution (**Table 3.2**) of class II antigen is restricted to cells involved in handling extra cellular antigen. These are dendritic cells, macrophages and B cells. Though various cytokines, especially interferon, enhances class II expression, its physiological significance is not clearly understood. The increased class II expression may increase the chance of successful presentation of antigenic peptide to helper T cells thereby increasing the level of the immune response to the foreign antigen.

Table 3.2: Distribution of class II antigens in different tissues.

Constitutive Expression	Inducible (IFN - Y)
B cells	Astrocytes / glial cells.
Erythroid precursors	Fibroblasts
Macrophages	Keratinocytes
Monocytes	Malanocytes NK cells T cells Thymic epithelium Vascular endothelium

Regulation of MHC Class II Expression

Expression of class II antigen is inducible by interferon (IFN - γ), which enhances mRNA and surface protein levels for HLA –DR, - DP and associated invariant chains in macrophages, endothelial cells, dermal fibroblasts, melanoma cells and leukaemia cells. DQ α and β genes are not inducible by IFN - γ.

The promoter region of HLA-DR contains three conserved sequences designated W, X and Y box. The X box consists of two tandem regulatory sequences called X1 and X2; the protein that binds to X2 is a member of the AP - 1 family of transcription factors. The W, X, Y boxes are occupied by proteins in all cells that express class II molecules.

An alteration in chromatin structure has been shown to correlate with class II gene expression and hypersensitive sites associated with looser chromatin structure have been identified after IFN γ treatment. DNA is modified by methylation of particular nucleotide sequences or cpG sites and the pattern of DNA methylation may be involved in controlling gene expression. In some situations demethylation of class II genes, appears to correlate with increased expression. It is important to note that this correlation is not absolute, since demethylation of at least one cpG site in the DRA gene blocks the expression of the gene.

IFN α regulatory sequence is located upstream from the class II genes in the same region of the W box. Cell type specific enhancers for DQA, DQB and DRA genes and transacting regulatory elements also play a role in control of gene expression. The "Bare Lymphocyte Syndrome" is a congenital immuno-deficiency disease characterised by the absence of cell surface expression of MHC molecules. The genetic defect in this syndrome does not lie within the MHC genes but is due to deletion of a transacting factor.

Organization of the Individual Class I and Class II MHC Genes

The general intron - exon organization of the class I and class II genes are similar; each structural domain of the polypeptide chain being encoded by a separate exon in genomic DNA (**Figure 3.5**).

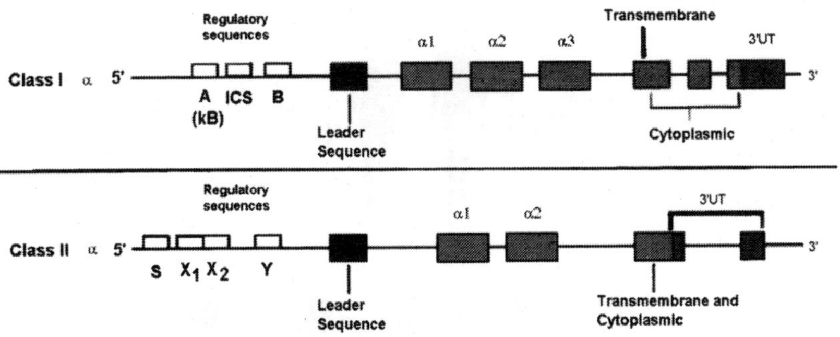

Figure 3.5 ***MHC gene structure*** *showing introns, exons, regulatory and leader sequences*

A typical human class I gene has eight exons, of which the first exon codes for the signal peptide, exons 2 to 4 encode the α1, α2 and α 3 domains and exon 5 codes for transmembrane (TM) region. Exons 6 and 7 code for the cytoplasmic domains, while exons 7 and 8 code for 3′ untranslated regions. The number of exons for the TM and cytoplasmic regions varies between different heavy chain genes. Nucleotide sequencing of HLA class I genes has shown that exons 2,3 and 4 are polymorphic with about 30% sequence variation between alleles.

In the class II genes, exon I codes for the leader or signal sequence that target the nascent proteins to the endoplasmic reticulum but is removed from the mature polypeptide chain. Exons 2 and 3 code for the extra cellular domains of the α and β chains, namely the N terminal domains α1 or β1, and the membrane proximal domains α2 or β 2 respectively. Exon 4 codes for the hydrophobic transmembrane region. In α chains, this exon also contains the sequence of nucleotides that codes for the cytoplasmic tail. In the β chain genes the amino acids that comprise the cytoplasmic tail are encoded by a separate exon. A final exon codes for the 3′ untranslated region of the α and β chains.

Inheritance of HLA

The MHC genes are co dominantly expressed and therefore each individual expresses two protein products of each locus at the cell surface. Because of close linkage between the genes with a recombination frequency of only 0.5% the genes at each locus on a single chromosome are usually inherited as a haplotype and each individual has 2 haplotypes (**Figure 3.6**). There is thus a 25% chance

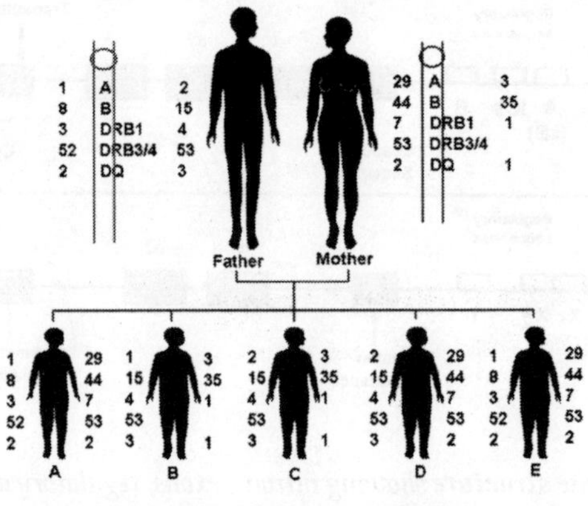

Figure 3.6 Schematic representation showing inheritance of HLA antigens

that two siblings will share both haplotypes and will be fully HLA compatible, 50% chance that they will share one haplotype, and 25% chance that they will share no haplotype and thus will be completely HLA incompatible.

Linkage disequilibrium

As a result of random mating in the population the frequency of an allele at one HLA locus being associated with an allele at second HLA locus, would be the product of the frequencies of each allele in the population. However, certain combinations of alleles are found with a frequency greater than expected. This phenomenon is termed as **linkage disequilibrium**, which may be defined as non-random association between alleles at different loci. Strong linkage disequilibrium in the HLA region has maintained many stable HLA haplotypes. HLA - A1 and HLA -B8 alleles are found in Caucasian population with a frequency of 0.161% and 0.104% respectively. The expected frequency with which A1-B8 haplotype should be found in this population is 0.161 x 0.104 = 0.0167% or 16.7 per thousand of population. However, it is found with a frequency of 59.2 per thousand, which is almost four times the expected frequency. The difference is denoted Δ (Δ = 0.0592 - 0.0167 = 0.0425 for A1 - B8) and the two loci are said to be in positive genetic disequilibrium.

Genetic linkage disequilibrium could have developed either due to selective advantage of a given haplotype in a population or recent admixture of two inbred population.

Generation of Polymorphism

Many mechanisms have been proposed for the generation of the extensive polymorphism seen at the MHC class I and class II loci. Overdominant selection or heterozygous advantage may be responsible for this polymorphism. Phenotypic selection pressure may give rise to point mutations, which may lead to allelic products. Multiple nucleotide substitutions in class II and I genes may also be responsible and are generally termed as gene conversion or copy substitution. The mechanism may be mediated through homologous region in H2 genes and could occur following duplication of a heteroduplex hybrid DNA formed as a recombination intermediate.

Large segments of the different alleles have conserved blocks of amino acids and observed variations can not be accounted for by recombination and normal rates of mutations alone. Thus, the role of evolution and selection for generation of polymorphism at this locus has to be conceded. There are various "hot spots" between DP and DR / DQ in the human D region. Other mechanisms such as gene duplication, deletion and differential RNA splicing could also generate diversity among class II molecules. In fact the close similarity between DRB1 and DRB3 has been attributed to recent gene duplication events. Another source of polymorphism of MHC class II products may be the alternative use of transcription signals in class II gene or of mutated translation signals in class II transcripts. All the class II genes do not use the same splicing and polyadenylation signals. Several cases of alternate splicing events have been described in the human system, for most of the genes. These events generally cover the exons encoding the cytoplasmic domain or the 3′ untranslated regions. Alternate use of different polyadenylation signals can generate two transcripts of different size in DR and DQ haplotypes. For other genes, alternate splicing generates 3′ untranslated and cytoplasmic exons of different sizes in different haplotypes of DQ and DP respectively. This polymorphism constitutes the major basis of tissue rejection in transplantation and is likely to play an important role in establishing the diversity of immune responsiveness.

Peptide MHC Interactions

What is the functional role of MHC molecules? The role of MHC proteins in antigen presentation was described more than a decade after the discovery of its role in transplantation.

In 1973, Hugh McDevitt and his colleagues demonstrated that susceptibility of mice to a particular pathogen was dependent on the allele present at one of the MHC loci. They found that lymphocyte choriomeningitis virus (LCMV) caused a lethal brain infection in mice that were either homozygous or heterozygous for the H-2^q alleles but did not cause infection in mice that were homozygous for the H-2^k allele at this locus.

These findings led Rolf Zinkegnagel and Peter Doherty to examine the role of cytotoxic T lymphocytes (CTLs) in the development of this disease. They tried to correlate the level of CTL activity with the severity of the disease in mice having different MHC genotypes (or haplotypes, as they are called). They evaluated CTL activity in spleen cells of mouse that had been infected with the LCM virus seven days earlier by measuring the lysis of target fibroblast (L cell) monolayer infected with LCMV. These targets were labelled with a radioisotope of chromium and target cell killing released the labelled chromium into the medium. Using this assay system, Zinkernagel and Doherty could demonstrate that spleen cells from infected H-2^k haplotype (e.g. CBA / H, AKR, and C3H/ Hej strains of mice) would only effectively kill targets, which were also of H-2^K type. Spleen cells from mice bearing H-2^b or H-2^d alleles were unable to lyse the infected H-2^k fibroblasts.

These results were reproducible for any combination of effector and target cells and these studies provided the first evidence that the MHC molecule on the surface of an infected cell restricts its interactions with T cells. This phenomenon is termed **MHC restriction.**

It was known that T cell stimulation occurs only in the presence of non-T cells. In mid-1970s and early 1980s, Alan Rosenthal, Emil Unanue and others demonstrated that these non-T cells were of monocytes/macrophages lineage and were involved in internalisation of antigens. These cells were termed as antigen presenting cells (APC) since they subjected the antigen to some alterations before T-cells could be stimulated.

These observations were made by incubating the APCs with antigen for one hour at either 4^0C (at which temperature no metabolic actively takes place) or 37^0C prior to adding them to T cell cultures. T-cell stimulation was measured by the incorporation of (^{3}H) thymidine into DNA synthesis that accompanies cell proliferation. Macrophages pre incubated with antigen at 37^0C were nearly 10 times more effective at stimulating T cells than those pre-incubated with antigen at 4^0C, suggesting that antigen processing requires active metabolic events. Treatment of cells with sodium azide, a cytochrome oxidase inhibitor, also inhibited the ability of APCs to stimulate T cells, indicating that antigen presentation requires metabolic energy.

Subsequent experiments by Kirk Ziegler and Emil Unanue demonstrated sequestration of extra cellular antigens into the macrophage occurred by endocytosis and delivery to the lysosomal compartment of cells. The endocytosis of antigens derived from the bacteria *Listeria monocytogenes* was not affected by chloroquine or ammonium chloride, both substances, which alter lysosomal pH.

However, both these substances markedly inhibited the processing of the antigen and its ability to stimulate T cells. These data were among the first to suggest that intracellular alterations of extra cellular antigens by active energy dependent processes in the lysosomal compartment is essential prior to presentation.

Other studies continued to implicate MHC molecules in the interaction between APCs and T cells. When Ziegler and Unanue treated macrophages with antibodies directed against MHC proteins, these antibodies had no effect on the uptake or catabolism of antigen, but markedly inhibited interaction of macrophages with T cells. The T cell-macrophage interaction was inhibited even if macrophages were exposed to antibodies before addition of antigen.

These and other studies indicated that the interaction between a T cell and macrophage was dependent on the recognition of two components on the surface of the antigen -presenting cell. Two models were proposed for antigen recognition by T cells; that T cells possess two distinct receptors, one for the antigen and another for the MHC protein or that a single T -cell receptor recognized both the MHC protein and the antigen peptide on the APC surface simultaneously.

X-ray crystallographic studies by Don Wiley and his colleagues in 1987 confirmed the earlier models based on the known structure of MHC molecules. They examined the structure of the extra cellular (soluble) portion of the MHC class I molecule (α1, α2, α3 and β2m) after removing the transmembrane anchor. The inner surfaces of α_1 and α_2 domains formed walls of a deep groove that was approximately 25Å long and 10Å wide. It is this groove that acts as the binding site for peptides produced by antigen processing in the cytoplasm. The sides of the antigen-binding pocket are lined by α helices from the α1 and α2 domains, and the bottom of the pocket is lined by β sheet that extends from the same domains across the mid-line. The helices are thought to form relatively flexible sidewalls enabling peptides of different sequences to bind within the groove.

Subsequent X-ray crystallographic studies described the manner in which peptides are positioned within the MHC antigen-binding pocket. In one of these studies, the spatial arrangement of several naturally processed peptides situated within the antigen-binding pocket of a single MHC class I molecule (HLA-B27) was determined. The backbone of all the peptides bound to HLA-B27 shared a single, extended conformation, running the length of the binding cleft. The

N-and C-termini of the peptides were precisely positioned by numerous hydrogen bonds at both ends of the cleft. The hydrogen bonds linked the peptide to a number of conserved residues in the MHC molecule that were part of both the sides and bottom of the binding groove.

Ian Wilson and colleagues have reported the X-ray crystallographic structure of a mouse MHC class I protein complexed with two peptides of different length. The overall structure of the mouse MHC protein is similar to that of the human MHC protein as shown in **Figures 3.7**. In both cases, the peptides are bound in an extended conformation, deep within the binding groove of the MHC molecule. This extended conformation allows for numerous interactions between the side chains of the MHC molecule and the backbone of the bound peptide. Because the MHC interacts primarily with the backbone of the peptide rather than its side chains, there are very few restrictions on the particular amino acid residues that can be present at various sites of the binding pocket. As a result, each MHC molecule can bind to a diverse array of antigenic peptides.

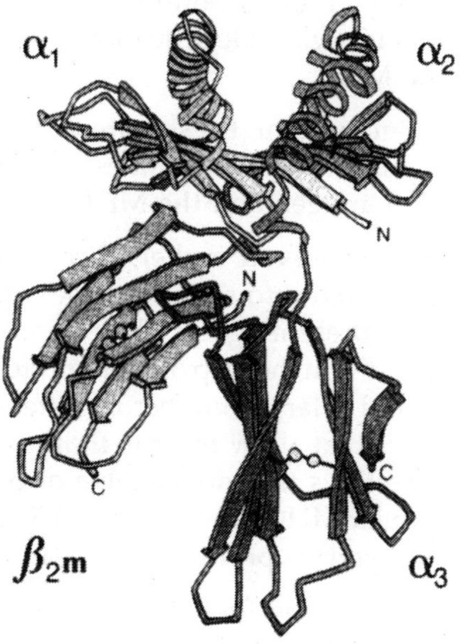

Figure 3.7 Schematic representation of a MHC class I molecule, in this case the human protein HLA-A2. *The molecule consists of two subunits; a heavy chain made up of three domains (α1, α2 and α3) and a β2m chain. The membrane-spanning portion of the heavy chain would connect with the polypeptide at the site marked C (for C-terminus of the remaining portion). Disulfide bonds are indicated as two connected spheres. The peptide-binding groove is shown at the top of the drawing, between the α helical segments of the α1 and α2 domains of the heavy chain.*

The MHC-peptide complex on the surface of the APC is only half the story of immunologic recognition; the other half is represented by the T-cell receptor (TCR) on the surface of the T cell. TCR can recognize both an MHC and bound peptides, but how this happened was not understood because of the difficulties in preparing protein crystals of the TCR that were suitable for X-ray crystallography.

The complementarity determining regions (CDRs) of the TCR interact with the α1 and α2 domains of the MHC as well as the exposed residues of the bound peptides. The central CDRs of the TCR, which, exhibit the greatest sequence variability, interact primarily with the centrally situated bound peptide, whereas the outer CDRs, which have a less variable sequence, interact most closely with the α helices of the MHC.

Because of these interactions, the TCR meets both of its recognition "responsibilities": it recognizes the bound peptide as a foreign antigen and the MHC as a self-protein.

As observed earlier, T cells are activated by antigen that has been broken into small peptides and displayed on the surface of an antigen-presenting cell. These peptide fragments are held at the surface of the APC in the grip of MHC proteins. Each species of MHC molecule can bind a large number of different peptides that share structural features that allow them to fit into its binding site. For example, all of the peptides that are capable of binding to a protein encoded by a particular MHC allele, such as HLA-B8, may contain a specific amino acid at certain position, which allows it to fit into the MHC groove.

Given the fact that each individual expresses a number of different MHC proteins, and each MHC variant may be able to bind large number of different peptides (**Figure 3.8**), a dendritic cell or macrophage should be able to display a vast array of peptides. In spite of this, some individuals are not able to present every possible peptide in an effective manner. This may be a major factor in determining differences in susceptibility with in a population to different infectious diseases. For example, the HLA-DRB1*1302 allele is associated with resistance to a certain type of malaria and hepatitis B infection.

The distribution of MHC alleles in a given population has been shaped by natural selection. Persons who possess MHC alleles that are best able to present the peptides of a particular infectious agent have highest probability of mounting an effective immune response against infection by that agent. Conversely, persons who lack these alleles are less likely to mount an effective immune response. In case of an infection that has fatal outcome, lack of an immune response will result in death and during major epidemics when large number of susceptible persons die, the concerned allele is lost. As a result, populations tend to be more resistant to diseases to which their ancestors have been routinely exposed. This could explain why Native American populations have been devastated by certain diseases, such as measles, that produce only mild symptoms in persons of European ancestry.

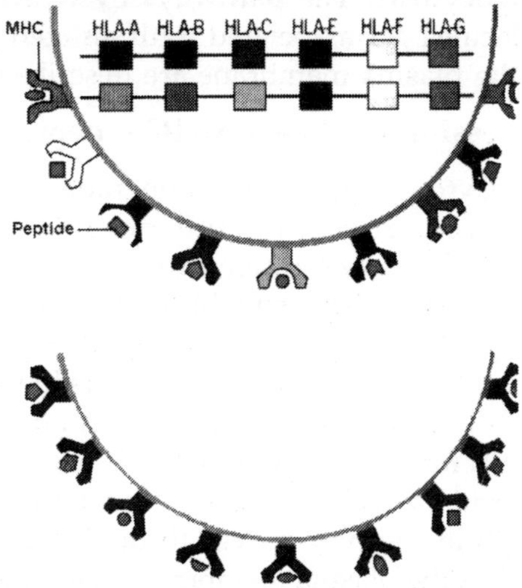

Figure 3.8 Human APCs can present large number of peptides. *(a) Schematic model of the variety of class I MHC molecules an individual might possess. This Class of MHC protein is encoded by six alleles, three of which are represented by a large number of alleles. This particular individual is heterozygous at the HLA-A, -B, and –C loci and homozygous at the HLA-E, -F and –G loci, which give them a total of 9 different MHC class I molecules. (b) Schematic model illustrating the variety of peptides that can be presented by the protein encoded by a single MHC allele.*

The entire process of T-cell mediated immunity is based on the differences in the small peptides derived from the proteins of a pathogen and those derived from the proteins of the host. Professional APCs (e.g. dendritic cells, macrophages, and B cells) are specialized for this function. When a T cell interacts with an APC, it does so by having its TCRs dock onto the MHC molecules projecting from the APCs surface. This interaction brings the TCR into an orientation that allows it to recognize the specific peptide displayed within a groove of the MHC molecule. The interaction between MHC proteins and TCRs is strengthened by additional contacts that form between cell -surface components, for example between CD4 or CD8 molecules on a T cell and MHC proteins on an APC.

The two classes of MHC molecules display antigens that originate from different sites within a cell, though some overlap has been reported. MHC class I molecules are predominantly responsible for displaying antigens that originate within the cytosol of a cell, that is, endogenous proteins. In contrast, MHC class II molecules primarily display fragments of exogenous antigens that are taken into a cell by phagocytosis. The pathways by which these two classes of MHC molecules get associated with antigen fragments and display them on the plasma membrane are described below.

Processing of class I MHC – peptide complexes.

Cytosolic antigens are degraded into short peptides by proteases that are part of a multimeric complex called proteasome. These proteases cleave antigen into fragments of approximately 8 to 10 residues, which are suitable for binding within an MHC class I groove. The peptides are then transported across the membrane of the rough endoplasmic reticulum and into its lumen by a dimeric protein called TAP (transporter associated with antigen processing). Once within the ER lumen, the peptide binds to a newly-synthesized MHC class I molecule, which is an integral protein of the ER membrane. The MHC -peptide complex moves through the biosynthetic pathway until it reaches the plasma membrane where the peptide is displayed (**Figure 3.9(a)**).

Processing of class II MHC-peptide complex: MHC class II molecules are synthesized as membrane proteins of the ER, and are joined non-covalently to a protein called Ii, which blocks the peptide-binding site of the MHC molecule. This prevents endogenous peptides from binding to the MHC groove. Following its synthesis, the MHC class II -Ii complex moves out of the ER along the biosynthetic pathway, directed by targeted sequences located within the cytoplasmic domain of Ii. MHC class I and II molecules are thought to separate from one another in the trans Golgi network (TGN), which is the primary sorting compartment along the biosynthetic pathway. While MHC class I peptide complex is directed toward the cell surface, an MHC class II-Ii complex is directed into an endosome or lysosome, where the Ii protein is digested by acid proteases. An MHC class II molecule is then free to bind to peptides digested from antigen that were taken into the cell and directed along the endocytic pathway. The MHC class II-peptide complex then moves to the plasma membrane where it is displayed (**Figure 3.9 (b)**).

Once on the surface of an APC, MHC molecules are involved in the interaction between the cells with different types of T cells. Cytotoxic T lymphocytes (CTLs) recognize their antigen in association with MHC class I molecules; they are said to be **MHC class I restricted**. Many cells of the body display fragments of their own normal proteins in association with their MHC class I molecules. However, very few

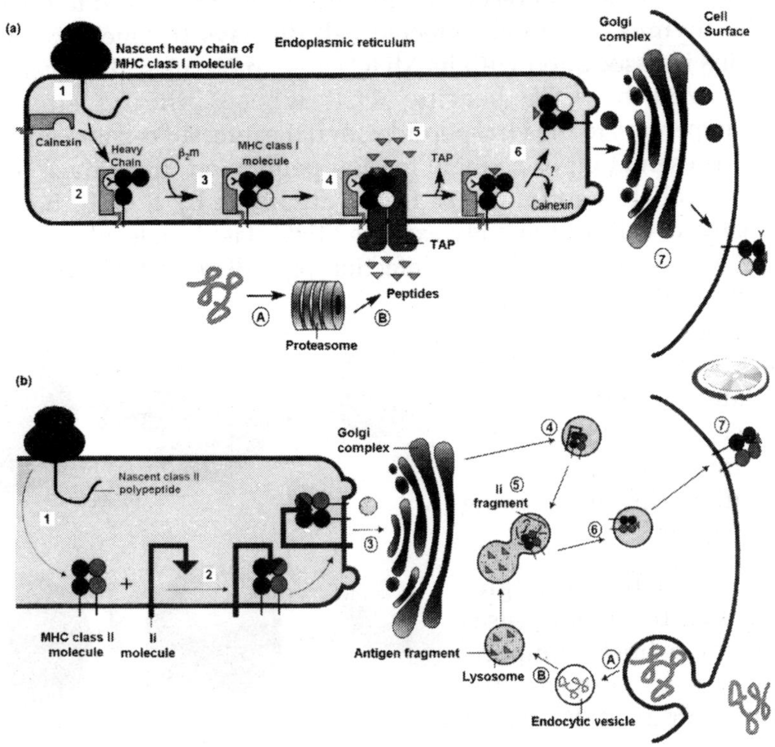

Figure 3.9 Processing pathways for antigens that become associated with MHC class I and class II molecules. *(a) Pathway for the assembly of an MHC class I-peptide complex. (1) heavy chain of the MHC protein is synthesized by a membrane-bound ribosome and translocated into the ER membrane. MHC heavy chain becomes associated with calnexin (2), a chaperone in the ER membrane, and the dimeric complex binds the invariant β_2 m chain (3). The MHC complex becomes associated with another ER membrane protein, TAP (4). Cytosolic antigens are taken into proteasomes (A) and degraded into small peptides (B). The peptides are transported into the ER lumen by the TAP protein, where they become bound within the groove of the MHC molecule (5). Calnexin and TAP dissociate from the MHC complex (6), which is transported along the biosynthetic/secretary pathway through the Golgi complex (7) to the plasma membrane, where it is ready to interact with the TCR of a CTL. (b) Pathway for the assembly of an MHC class II-peptide complex. MHC protein is synthesized by a membrane-bound ribosome and translocated into the ER membrane (1) where it becomes associated with Ii (2), a protein that blocks the MHC-peptide-binding site. The MHC-Ii complex passes through the Golgi complex (3) and into a transport vesicle (4). Meanwhile, an extra cellular protein antigen is taken into the APC by endocytosis (A) and delivered to a lysosome (B), where the antigen is fragmented into peptides. The lysosome containing antigenic fragments fuses with the transport vesicle containing the MHC – Ii complex (5), leading to the degradation of the liprotein and association between the antigenic peptide fragment and the MHC class II molecule (6). The MHC-peptide complex is transported to the plasma membrane (7) where it is ready to interact with the TCR of a T_H cell.*

T cells with TCRs that recognize auto antigens are allowed to develop in the thymus. When an infected cell displays fragments of viral proteins in association with its MHC class I molecules, these cells are recognized by CTL bearing TCR whose binding sites are complementary to the viral peptides, and the infected cell is destroyed. T cell presentation of a single foreign peptide on the surface of the cell is probably sufficient to invite an attack by a CTL. Because virtually all cells of the body express MHC class I molecules on their surface, CTLs can combat an infection regardless of the type of cell affected. CTLs may also recognize and destroy cells that display abnormal (mutated) proteins on their surface, which could play a role in the elimination of potentially life threatening tumour cells.

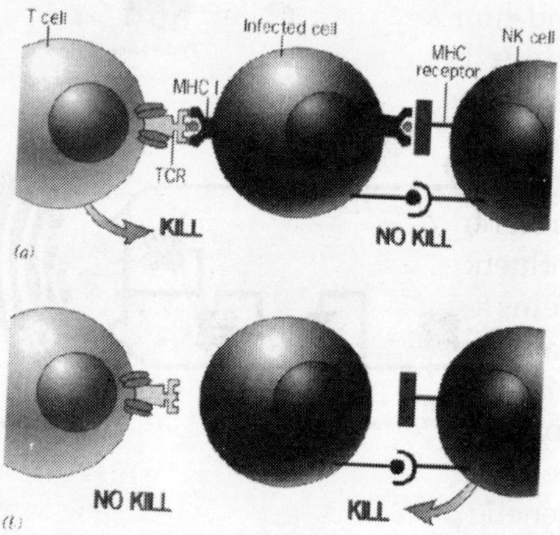

Figure 3.10 ***MHC proteins and viral infections.*** *(a) Cytotoxic T lymphocytes (CTLs) recognize infected cells by the viral peptides presented in combination with an MHC molecule. Natural killer cells, on the other hand do not kill a cell displaying MHC molecule even if it is infected.(b) Some viruses evade CTL destruction by causing the host cell host cell to stop displaying MHC molecules. The absence of MHC proteins on a cell surface makes it a target for NK cell*

One of the strategies employed by some viruses to evade a host's immune system is to suppress the expression of MHC class I molecules. Some metastatic tumour cells lose MHC expression marking the cell "invisible" to cytotoxic T cells. NK cells normally do not lyse cells that bear MHC class I molecules . But, cells, which lose MHC class I expression either due to viral infection or malignant transformation become NK cell targets (**Figure 3.10**).

Helper T cells recognize antigen in association with MHC class II molecules; and are therefore activated by exogenous (i.e., extra cellular) antigens (**Figure 3.9**), such as those that occur as part of bacterial cell walls or bacterial toxins. MHC class II molecules are present predominantly on B cells, dendritic cells, and macrophages. Macrophages and dendritic cells, which ingest foreign, extra cellular materials, present the fragments to helper T cells. Helper T cells that are activated in this way can then stimulate B cells to produce antibodies that bind to the exogenous antigen wherever it is located in the body.

If the primary function of MHC molecules is antigen presentation, why are the genetic loci encoding these proteins so highly polymorphic? MHC polymorphism gives members of a population an individuality by which they can be distinguished from other members. A series of fascinating behavioural studies in both mice and humans have indicated that a characteristic difference in body odour between persons (or mice of different inbred strains) is associated with difference in specific MHC alleles. These differences may arise from soluble MHC proteins excreted in sweat and the effects of these proteins on the growth of bacterial flora. In fact mating preferences in mice and humans have been shown to be strongly influenced by MHC genotypes. When women were asked to select items with a pleasurable aroma from among sweaty clothing that had been worn by different men, they selected items from men with MHC that were most different from their own. These observations could have a wider implication in evolutionary terms as mating between individuals with different MHC alleles would maintain wider genetic pool. As far as the MHC is concerned this would also give the population ability to display larger range of peptides.

Distinguishing self from Non-self

T cells gain their identity in the thymus. When an immature T cell migrates to the thymus from the bone marrow, it lacks the cell surface proteins that mediate T -cell function, most notably its TCRs. Once a T cell undergoes DNA rearrangements that enable it to produce a specific TCR, the cell is subjected to a complex screening process in the thymus that selects cells having potentially useful T cell receptors. T cells who's TCRs have a high affinity for peptides derived from the body's own proteins are destroyed (**Figure 3.11a**). This process of **negative selection** greatly reduces the likelihood that the immune system will attack its own tissues.

The generation of T cells requires more than negative selection. When a TCR interacts with a foreign peptide on the surface of an APC, it must recognize both the peptide and the MHC molecule holding that peptide. Consequently, T cells whose TCRs do not recognize self-MHC molecules are of little value. The immune system screens out such cells by requiring that T cells recognize self-MHC self-peptide complexes with low affinity. T cells who's TCRs are unable to recognize self-MHC complexes undergo apoptosis in the thymus, a process referred to as "death by neglect" (**Figure 3.11 b**). In contrast, T cells with TCRs that exhibit weak (low affinity) recognition toward self-MHC -self-peptide complexes are stimulated to remain alive but are not activated (**Figure 3.11 c**). This process of selective survival is referred to as **positive selection.**

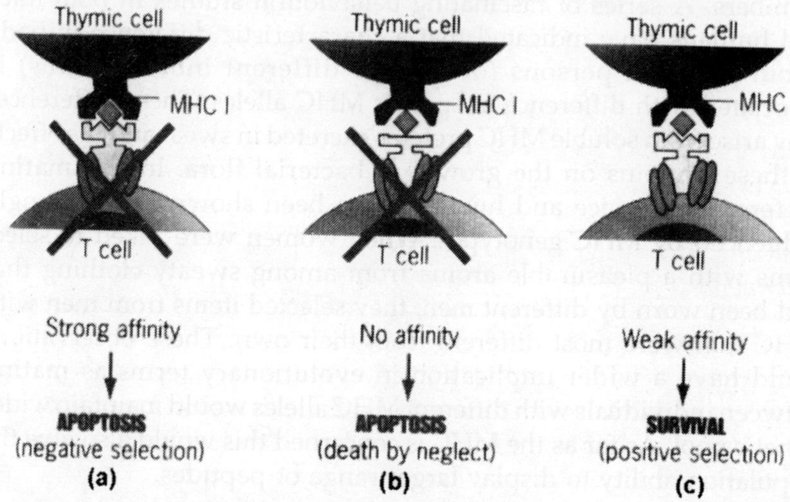

Figure 3.11 Determining the fate of a newly formed T cell in the thymus. *A screening process takes place in the thymus that selects for T cells with appropriate TCRs. (a) Those T cells whose TCR exhibits strong affinity for MHC molecules bearing self-peptides are eliminated by apoptosis (negative selection). (b) Those T cells whose TCR fails to recognize MHC molecules bearing self-peptides also die by apoptosis (death by neglect). (c) In contrast, those T cells whose TCR exhibits weak affinity for MHC molecules bearing self-peptides survive (positive selection) and ultimately leave the thymus to constitute the peripheral T cell population of the body.*

Recognition of class I self-MHC complexes commits an immature T cell to differentiate into a CD8+ killer T cell, whereas recognition of class II self-MHC complexes commits an immature T cell to differentiate into a CD4+ helper T cell. It is estimated that less than five percent of T cells that mature in the thymus survive these negative and positive screening events. T cells that do survive leave the thymus and migrate to the peripheral lymphoid tissues to deal with any foreign antigen.

Evolution

MHC polymorphism has been created through the process of evolution. Some of the new alleles may have evolved as result of point mutations but many have arisen from combination of sequences from different alleles either by genetic recombination or by gene conversion in which one sequence is replaced, in part by another from a homologous gene.

Comparison of the sequences of different MHC alleles, show that differences in alleles involve clusters of several amino acids that require multiple nucleotide changes in a contiguous stretch of the gene. Even more significantly, the same sequences are found within other MHC genes of the same chromosome, a prerequisite for gene conversion. Recombination between alleles at the same locus may, however, have been more important than gene conversion in generating MHC polymorphism. A comparison of sequences of MHC allele shows that many different alleles could represent recombination events between a relatively small set of hypothetical ancestral alleles.

Selection has played an important role in establishing HLA polymorphism. New polymorphism might have occurred as a result of point mutation i.e. silent substitutions, which simply change the codon but leave the amino acid sequence unchanged. Replacement substitutions occur within the MHC at a higher frequency relative to silent substitutions than would be expected providing evidence that polymorphism has been actively selected for in the evolution of the MHC.

MHC and Disease Association

With the intimate involvement of the MHC in initiation of the immune response, it is not surprising that association studies have shown several diseases caused by immunological dysfunction to be associated with specific MHC alleles. Some non-immunological diseases have also been found to be associated with this chromosomal region since many genes without immunological functions are also present in the region. Both these expectations have been borne out by disease association studies in humans and animals. HLA alleles have been found to occur more frequently in patients with some diseases, than they do in groups of healthy individuals a phenomenon termed "disease association". This association is 100%, in the case of narcolepsy (a disease characterized by a sudden, uncontrollable compulsion to sleep during waking hours) with virtually all patients of narcolepsy having HLA -DRB1*02 allele. However, all individuals who have the HLA-DRB1*02 allele do not suffer from narcolepsy. All other HLA disease associations are of a statistical nature, the association being stronger in some than others. For example, ankylosing spondylitis, a disease characterized by inflammation and stiffening of the vertebrae, ossification of the spinal ligaments and ensuing spinal deformities, occurs in Caucasians with a frequency of 0.5-4 per 1000 males and 0.05-0.5 per 1000 females. Approximately 90% of the Caucasians afflicted with this disease carry the HLA -B* 27 allele. Since the allele is present in only 9% of healthy individuals, its frequency in patients with ankylosing spondylitis is increased 10 times, so that individuals carrying the B*27 allele are clearly more susceptible to the disease than individuals carrying other HLA alleles. In this case, the association between HLA and the disease is indisputable and apparent even in a small sample of patients. In other cases, however, the association may be so weak that very large groups of patients must be tested to prove its statistical significance. For example, Hodgkin's disease, characterized by the neoplastic transformation of B-lymphocytes in a specific stage of their development, occurs with a frequency of about 4 in 100000. About 40% of patients with this disease carry the HLA-A*01 allele in comparison to about 32% in a control population.

Some well-established HLA-disease associations are listed in **Table 3.2**. Many other associations have been reported, but they are weak and their significance is not established. For example, myasthenia gravis, a disease characterized by muscular weakness that may progress to paralysis, was originally found to be associated with HLA-B*08 but later a stronger association with HLA-DRB1*03 was recognized (the two alleles often occur together in a haplotype). Although the diseases listed in **Table 3.2** are of widely varying aetiology, a number of them have certain features in common. With the exception of juvenile diabetes, they are all rare, occur after the age of reproduction and have little or no effect on viability or fertility. In most cases, the HLA genes are associated with susceptibility rather than resistance to the disease and the associations are often with genes that are found at a relatively high frequency in the population. Several of the diseases have an autoimmune character ie they are characterized by the presence of antibodies or T lymphocytes specific for self-components. Many of the diseases affect joints (ankylosing spondylitis,

Reiter's syndrome, rheumatoid arthritis, systemic lupus erythematosus, sicca syndrome), endocrine glands (Grave's disease and thyroiditis affect the thyroid gland) or skin (Behcet's disease, psoriasis, dermatitis herpetiformis and pemphigus).

Congenital adrenal hyperplasia is the only disease in the list for which the cause is known. It is associated with HLA-B*47, and is caused by a mutation in the gene coding for the enzyme 21-hydroxylase, one of the enzymes involved in biosynthesis of corticosteroids. This defect leads to an overgrowth of the adrenal gland and excessive secretion of adrenal hormones. The gene for the enzyme is located within the HLA complex and the reason for the association is that the mutation that inactivate the 21-hydroxylase gene apparently occurred in an individual carrying the HLA-B*47 allele and was then passed onto the progeny, together with this allele. Many of the patients with this form of adrenal hyperplasia are probably descendants of this mutant individual. The reasons for the association of the other diseases with HLA are, however, not known. Evidence for the involvement of HLA genes themselves is accumulating. It has been demonstrated, for example, that pathological conditions resembling ankylosing spondylitis develop in some mice and rats made transgenic for HLA-B*27 gene. There is also evidence that two of several genes controlling susceptibility to insulin-dependent diabetes mellitus (IDDM) are those specifying the HLA-DQ molecules. According to one hypothesis, individuals with Arg 52 in both of their DQ α chains and lacking Asp 57 in both of their DQ β chains have the highest susceptibility to IDDM.

Several hypotheses have been put forward to explain how HLA genes might be involved in disease susceptibility. The **molecular mimicry** hypothesis, suggests that antigens carried by a particular pathogen and certain HLA allomorph (product of an allelic gene) share structural similarities. Individuals carrying this allomorph are immunologically unresponsive to the pathogen and hence susceptible to the disease caused by the pathogen. For example, one of the antigens of *Klebsiella* has been shown to resemble HLA -B27 and the pathogen is thought to be responsible for initiating the process leading to ankylosing spondylitis. The **receptor hypothesis** proposes that HLA molecules are required for the expression of receptors for pathogens, hormones or drugs. For example, patients with celiac disease (a disease characterized by an intolerance to gluten, a protein present in wheat, rye, oats and barley) possess a receptor for gluten and that HLA-B*8, with which the disease is associated, is required for the expression of this receptor. The **Ir-gene hypothesis** suggests that the association between HLA and disease arises because certain individuals are unresponsive to certain critical epitopes of the pathogen that is presumably responsible for the disease. The **altered-self hypothesis,** proposes that agent responsible for the disease alters molecules of the patient's body, which are then regarded as foreign by the body's immune system and consequently become the target of an immune attack. While it is possible that each of these proposed explanations apply to a different disease, there is little evidence to support any of them.

Since the function of MHC molecules is the presentation of antigenic peptides to T cells, one might have expected that infectious diseases would show strong associations with HLA alleles. Whenever an MHC molecule fails to present a parasite-derived peptide, the concerned individual (lacking other MHC molecules capable of presenting the peptide) should become susceptible to this parasite. The association of HLA-DR2 with lepromatous leprosy, a severe form of infection by Mycobacterium leprae that spreads uncontrolled by the immune system, is relatively weak. (Here, 'DR2' stands for serologically defined antigen controlled by the HLA-DRB1 locus). The association of HLA-B53 (again an antigen designation) with a severe form of malaria has been observed in some African populations but could not be confirmed in others. The reasons for general lack of association between HLA and infectious diseases are unclear. One possibility is that selection has eliminated susceptible alleles from the population and that the alleles now present in the population provide an equal degree of protection. It should also be kept in mind that the antigens of infectious diseases are complex and hence every infected individual is able to respond to at least some of their antigenic components. One could also argue that most infectious diseases do not show association with HLA because their causative agents have become so destructive that the MHC-based immune system can no longer deal with them efficiently. However, none of the speculations have been substantiated and it is still not evident why the MHC appears to have so little influence on susceptibility to infectious diseases.

Table 3.2 Association between HLA and disease

Disease	HLA marker	Frequency (%)		Relative risk
		Patients	Controls	
Rheumatological disorders				
Ankylosing spondylitis	B27	90	9	87.4
Reactive arthropathy including Reiter's	B27	79	9	37.0
disease	B27	52	9	10.4
Acute iridocyclitis	DR4	69	34	4.2
Rheumatoid arthritis	B27	25	9	3.2
Juvenile chronic arthritis	DRB1*11	55	25	3.8
Pauciarticular early-onset juvenile chronic	DRB1*08	38	7	9.9
arthritis	DPB1*			
	B51	29	10	3.8
Behcet's disease	DR3	66	25	5.8
Systemic lupus erythematosus (SLE)	B8	48	25	2.8
Organ-specific autoimmune endocrine				
diseases	DR3	52	25	3.3
Insulin-dependent diabetes mellitus (IDDM)	DQB1*0201	66	45	2.4
	DR4	77	34	6.4
	DQB1*0302	60	14	9.5
	DR2	7	28	0.19
	DRB1*0501			
	DRB5*0101			
	DQB1*0602	3	17	0.15

Disease	HLA marker	Frequency (%)		Relative risk
		Patients	Controls	
Idiopathic Addison's disease	DR3	67	25	6.3
Grave's disease	DR3	55	25	3.7
Hashimoto's thyroiditis	DR11	28	11	3.2
Postpartum thyroiditis	DR4	73	34	5.3
Other organ-specific autoimmune disorders				
Celiac disease	DR3	78	25	10.8
	DQB1*0201			
	DQA1*0501			
	DR7,11	0-44	1-4	6-10
	DR7,DQB1*0201			
	DR11,DQA1*0501			
Dermatitis herpetiformis	DR3	84	25	15.9
Sicca syndrome	DR3	76	25	9.7
Myasthenia gravis	DR3	45	25	2.5
	B8	53	25	3.4
Idiopathic membranous nephropathy	DR3	80	25	12.0
Good pasture's syndrome	DR2	86	28	15.9
Multiple sclerosis	DR2	61	28	4.1
	DRB1*1501			
	DR5*0101			
	DQB1*0602			
Optic neuritis	DR2	48	28	2.4
Narcolepsy	DR2	99	28	169
Pemphigus vulgaris (Jewish population)	DR4	87	32	14.4

Disease	HLA marker	Frequency (%)		Relative risk
		Patients	Controls	
Pernicious anaemia	DR5	40	11	5.4
Alopecia areata	DR4 & DR5			
	DQB1*0301	65	23	6.1
Psoriasis vulgaris	Cw6	87	33	13.3
Birdshot retinochoriopathy	A29		8	109
Sub acute thyroiditis	B35	70	15	13.7
Infections				
Lepromatous leprosy	DR2			
Severe malaria	B53	17	25	0.61
Other immune disorders				
Immunization against HPA-1a platelet-	DR3	97	25	113.0
specific alloantigen	DR52			
IgA deficiency	DR3	50	25	3.0
Also associated with DR7				
C2 deficiency	Associated with the A25, B18, DR2 haplotype			
C4 deficiency	Associated with the B8, DR3 haplotype			
Malignancies				
Hodgkin's lymphoma	A1	40	32	1.4
Acute lymphatic leukaemia	A2	62	53	1.4
Non-immune disorders				
21-hydroxylase deficiency	B47	6	0.4	15.4
Idiopathic haemochromatosis	A3	76	28	8.8

Explanations in HLA marker column, symbols with asterisks indicate alleles; symbols without asterisks indicate serologically defined antigens. For every disease, the markers showing the strongest associations are given. In many cases in which it is difficult to decide whether HLA-DR or DQ markers are responsible for the association both markers are given.

The strength of HLA-disease association is calculated from 2 x 2 tables:

	Number of individuals	
	Marker present	Marker absent
Patients	a	b
Controls	c	d

Frequency of marker in patients: fp = a / a + b

Frequency of marker in controls: fc = c / c + d

Absolute risk AR = F (fp / fc)

Relative risk: RR = ad / bc

Relative risk (RR) indicates how many times more frequently a disease occurs in individuals with the HLA marker compared with the frequency of the disease in individuals without this marker. For positive association (i.e. when the HLA marker is more frequent in patients than in the control sample), RR is > 1.0; for negative associations it is < 1.0; and when there is no association, RR = 1.

An aetiological fraction (AF) indicates how much the HLA marker in question 'contributes' to the disease at the population level. It can be estimated for positive association only, in which case it can range from zero (no association) to 1.0 (absolute association). For negative associations, a preventive fraction (PF) can be estimated.

	No. of individuals (%) HLA – B27-postive	HLA-B27-negative	Total
Patients with ankylosing spondylitis	108 (90%)	12	120
Controls	311 (9.4%)	2990	3301
Total	419	3002	3421

RR = (108) (2990) / (12) (311) = 86.5, AR = (0.003) (0.90/0.0094) = 0.0286, EF = [(86.5-1) / (86.5)] (0.90) = 0.89, i.e. 89%

Techniques to detect HLA antigens:

In all types of transplantations (solid organ or bone marrow), graft from a donor, is recognized as foreign and has a potential to elicit an immune response in the recipient. Therefore, for a successful transplantation, matching of the donor and recipient MHC molecules is important. In case of incompatibility, two situations can occur:

1. Host versus graft reaction in which the MHC molecule of the host recognize the graft as foreign and mount an immune response against the graft leading to graft rejection. This is the usual response to solid organ grafts.
2. Graft versus host disease (GVHD): In the case of bone marrow transplantation, the donor tissue is bone marrow, which carries immuno-competent cells. Since these are infused into a recipient whose own haematopoietic cells have been totally depleted, donor cells mount an immune response against the recipient antigens leading to acute GVHD. This syndrome is characterized by skin rash, diarrhoea and jaundice or chronic GVHD, which results in susceptibility to infection.

To select fully / maximally matched donors, recipient and the donor need to be matched for HLA antigens. This can be done either by serological or molecular methods.

Serological typing by Complement dependent lymphocyto-toxicity

The micro-lympho-cyto-toxicity test is universally employed for serological typing of HLA-A, -B, -C, -DR, and -DQ loci. Peripheral blood mononuclear cells (PBMC) consisting of approximately 80% T and 10-15% B cell in healthy subjects are generally used as targets for typing for HLA class I locus. Since only antigen presenting cells express class II loci, cells enriched for B-lymphocytes are used for HLA-DR and -DQ typing.

HLA antigens are tested with a set of HLA specific alloantibodies which are either polyclonal antisera or monoclonal antibodies each selected to react with one (monospecific) or sometimes two or more (polyspecific/ multispecific) HLA antigens. For complete HLA typing the antibody panel should represent as many of the various HLA antigens as possible. Reagents for HLA typing should ideally be selected to give strong reliable positive reading when the antigen is present and negative reading when it is absent. The interpretation of a weak or doubtful positive reaction requires knowledge of the reactivity and specificity of the antisera. Since differences in antigen frequencies are well known in the different ethnic groups, ethnicity of the subjects must be taken into account when selecting antibody panels for HLA typing.

Principle of lymphocytotoxicity

In microlymphocytotoxicity assay, a specific antibody reacts with HLA antigen on the cell membrane and lyse cells in the presence of complement. In the classical microlymphocytotoxicity assay, cells are incubated with sera, for 30 minutes followed by further 60 minutes incubation after adding rabbit complement. During cell lysis, cells become permeable to vital dyes such as Eosin-Y, Trypan blue, or fluorochromes, such as Ethidium Bromide and Propidium Iodide. Thus cell death (positive test) indicates the presence of the antigen against which the reagent antibody is directed, while cell viability (negative test) indicates absence of the specific antigen. The degree of damage to lymphocytes can also be assessed by the loss of fluorescent diacetate or Acridine orange from pre labelled viable lymphocytes. This is illustrated in **Figure 3.12**

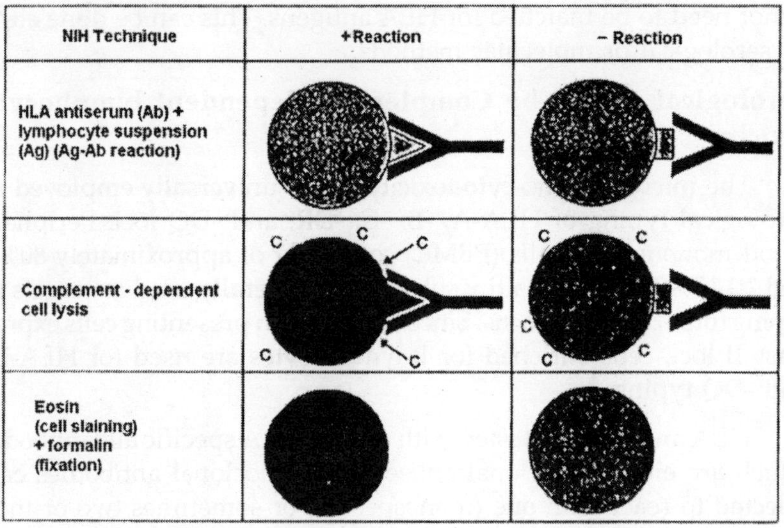

Figure 3.12 Principle of lymphocytotoxicity test

Lymphocytotoxic crossmatch

In solid organ transplantation, pre-existing antibodies against donor cells are deleterious and can result in hyper acute rejection. To prevent this, pre transplant cross match test is performed. This test detects anti-lymphocyte antibodies (usually anti- HLA) in recipient serum. If the serum has anti donor antibodies, these bind to the cell surface of donor lymphocytes killed by addition of complement (rabbit serum). The presence of dead cells is detected by dye exclusion methods as has been discussed in HLA typing section.

The antibodies detected by complement-dependent cytotoxicity may be against HLA class I or II antigens. Anti-class I (Anti-HLA-A, B, C, W4/W6) antibodies are detected by using purified suspension of peripheral blood lymphocytes (PBL) as target cells. Temperature variations in the first stage (serum and cell incubation) are used to enhance reactivity and also determine temperature-dependent class I antibodies.

Anti-class II (Anti-HLA-DR) antibodies can be detected by using purified suspension of B cells as target cells. Variation in temperatures (22^0, 37^0, 15^0C) during the 60 minutes first stage of the microlymphocytotoxicity assay is used for proper antigen antibody binding and also to determine temperature dependency.

The sensitivity of the reaction depend upon

1. Quantity of antigen (i.e. number of cells)
2. Titre of the antiserum
3. Conditions of the 2nd stage complement phase (i.e. activity of the complement, presence of complement inhibitors etc).

Molecular HLA Typing

Molecular HLA typing techniques are robust and more reliable as compared to serology and are widely used. Specified stretches of DNA, which reflect the allelilic differences between the HLA antigens, are amplified by PCR. The nomenclature for the alleles of the HLA system comprise of (i) two letter gene name (ii) the specific locus name (iii) two digits to define the allele group (iv) two digits for the subtype with an asterik between ii and iii. For example, serologically defined specificity DR7 will be written as DRβ1*07, since this is the allele of the β1 locus of DR gene. The subtypes of DRβ1*07 will be designated as DRβ1*0701, or DRβ1*0704. This more extensive and precise characterization of HLA alleles has been invaluable in trying to match transplant donors and recipients. Identification of the individual alleles may be done by one of the following methods: Probe hybridisation, Sequence Specific Typing or Direct Sequencing.

Sequence Specific Oligonucleotide Probe (SSOP) typing

Probe hybridisation technique is based upon amplification of a locus specific target DNA sequence, which is generally immobilized onto a support membrane (dot blotting). The polymorphism in the immobilized amplified DNA are subsequently detected by using specific single stranded DNA probes specific for each allele at that locus in combination with highly stringent washes to remove non-specifically bound probe. PCR-SSOP (Sequence Specific Oligonucleotide Probes) is a generally accurate method of genotyping.

The SSOP technique involves the following steps:

1. Amplification of specific region of the locus to be typed
2. Dot-blotting of the amplified DNA
3. Labelling of oligonucleotide probes
4. Hybridization of blotted product with labelled probe
5. Detection of the hybridized probes
6. Deprobing / rehybridisation of probes
7. Interpretation of signals

For generic amplification, primers are designed to amplify all known alleles at the a given locus and are complementary to sequences shared by all the alleles (Figure 3.13). An example is shown in **Figure 3.14** for the generic amplification of the exon II of the DRβ1 gene. For the specific amplifications for sub-typing, primers having complementarity to a sub set of alleles (e.g. all DR2 alleles) only are chosen.

The amplified PCR product is applied to nylon membranes and cross- linked to bind DNA to the nylon membrane.

The probes used for detection may be DIG-labelled, ^{32}P labelled or biotin labelled. The membrane is hybridised using the labelled probe under stringent condition to ensure high specificity.

The hybridized probes are detected using the appropriate method (**Figure 3.15**).

For every SSO, known positive and negative controls should be included. Any spot stronger than the negative control is considered positive. When the difference between positive and negative signals are very small, an additional critical wash with a slightly increased temperature is carried out, so as to make the difference between the signals of positive and negative controls more meaningful. In general,

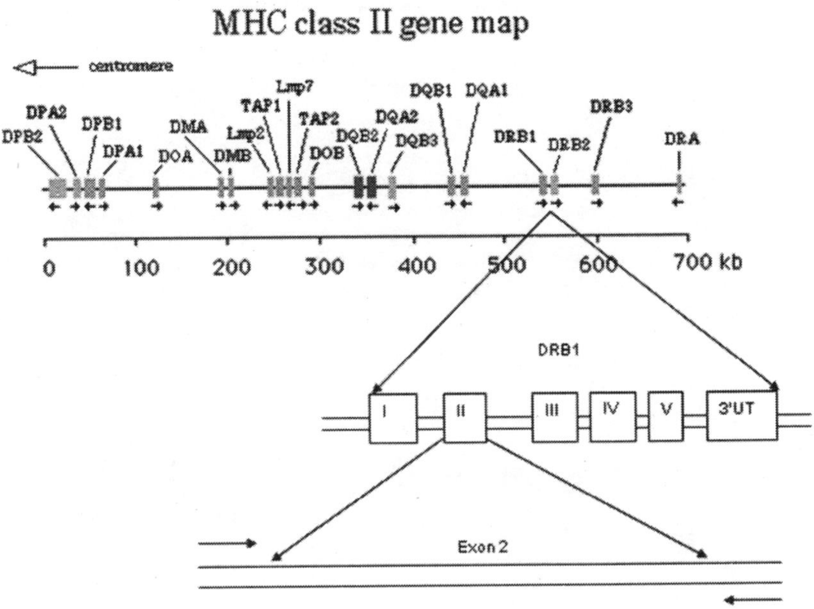

Figure 3.13. Schematic representation of primer designing for generic amplification of exon2 of DRB1 gene.

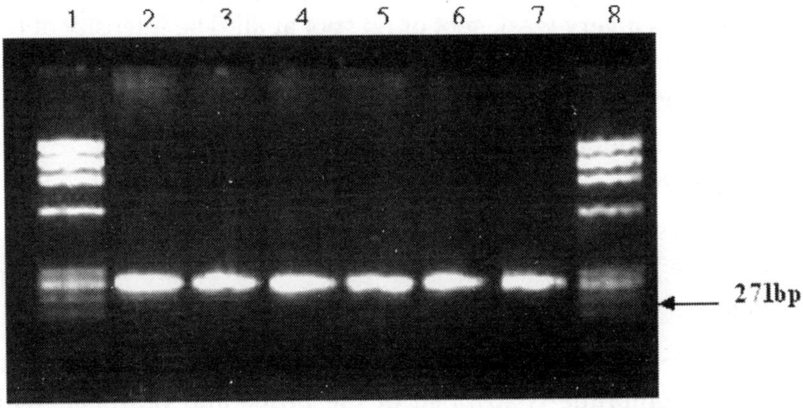

Figure 3.14 Product of DRB1 gene exon 2 amplification using flanking primers. *Lane 1 & 8 shows the molecular weight marker Hae III digest of ϕ X 174 DNA. Lane 2-7 shows amplified product of DRB1 exon 2 using primers 2DRBAMPA & 2DRBAMP-B from 6 samples.*

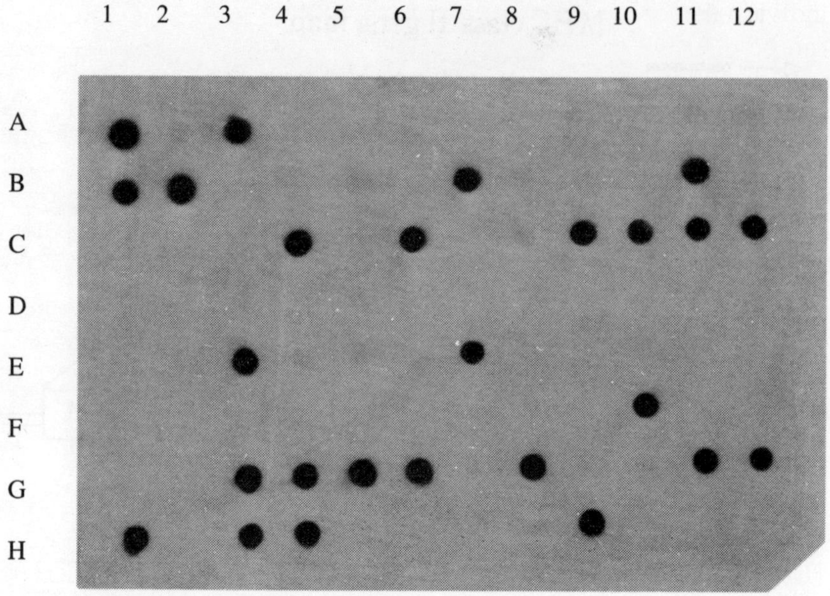

Figure 3.15 A representative autoradiograms showing the positive control and test positive dots. *Position G3-6/9/11-12, H1/3-4/9 have positive controls. Positions G1/27/9/10, H2/5-8/10-12 are negative controls.*

for each SSO, the exposure time is such that the negative control on the film shows a very weak spot or no spot at all. The intensity of the spots is represented by grades as follows:

0 Not spotted
1 Negative (definite)
2 Negative (probably)
4 Indefinite
6 Positive (probably)
8 Positive (definite)
9 Positive (definite, more than double intensity).

The membrane is stripped of the probe and reprobed with specificity. Individual probes are scored separately. When probing for all essential probes required for the assignment of the particular alleles are completed, the assignment of HLA class II alleles are performed using the interpretation tables.

This method is especially suitable for typing large number

individuals since generic amplification products from many individual can be detected together on a single membrane (**Figure3.16**)

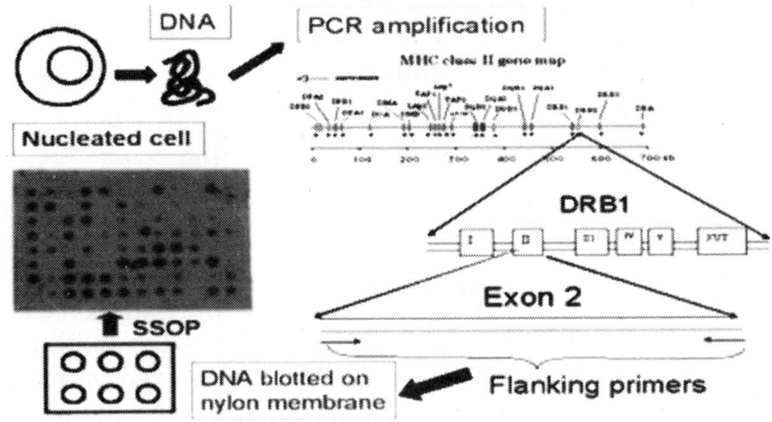

Figure 3.16 Sequence specific oligonucleotide probe-based typing

Reverse dot blot method

In reverse dot blot analysis, membrane strip carries specific oligonucleotide probes for all the alleles of a locus. (**Figure 3.17**). The genomic DNA is amplified by PCR using a locus-specific primer pair spanning several exons. The PCR product is then hybridised with this membrane. Positive signals are recorded and the allele assignment is done.

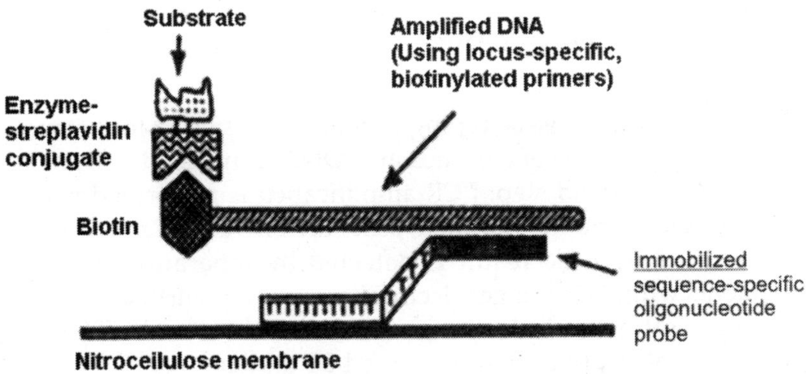

Figure 3.17 PCR-based HLA genotyping with the reverse dot blot format

The relatively short turnover time of 4 hr makes PCR methods suitable for HLA typing for cadaveric organ transplantation. In most instances, homozygotes and heterozygotes can be easily distinguished. As the DNA sequences of new HLA specificities are identified, the sequence data can be accessed via computerized networks. Suitable oligonucleotide primers or probes can be designed and rapidly incorporated in the clinical HLA laboratory.

SSP or Sequence Specific primer based typing

Amplification of alleles at a locus using allele specific primers is used quite extensively. This method is often referred to as sequence specific primer or amplification refractory mutation system (ARMS) typing. While SSP and ARMS have a slightly different context, the principle is very much the same. In these systems, the amplification process is performed using one primer that is conserved across many alleles and a second primer specific for one base pair (bp) difference. That is not to say primer is 1bp in length, but, that the 3′ end of the primer is specific for a complementary base in the target. The 3′ end of a primer must be firmly bound to the target for the polymerase to extend the strand. The detection system employed for the product of PCR is usually on an agrose gel electrophoresis. Most of the SSP kits available have multiple primer sets in separate reactions to type for many alleles at a locus or loci. It is essential to optimise melting temperature of the primer to allow thermal cycling of all reactions in a single PCR reaction. Testing using SSP and ARMS can yield low, medium or high resolution results, depending on the primer sets used.

Figure 3.19 presents a diagrammatic view of the location of HLA-DR-specific primers in relation to polymorphic and constant gene regions.

The PCR-SSP tissue typing procedure consists of four steps. The first step is extraction of chromosomal DNA from blood, cells or crude tissue. In the second step, PCR amplification is performed using the extracted DNA and the supplied, primers from recognized companies. In the third step the result is detected by separation of the PCR fragments with Agarose gel electrophoresis. The fourth and final step is documentation and interpretation of the result. The whole procedure takes about two hours and 30 minutes.

HLA typing by DNA sequencing

Sequence based HLA typing (SBT) is the only method to define

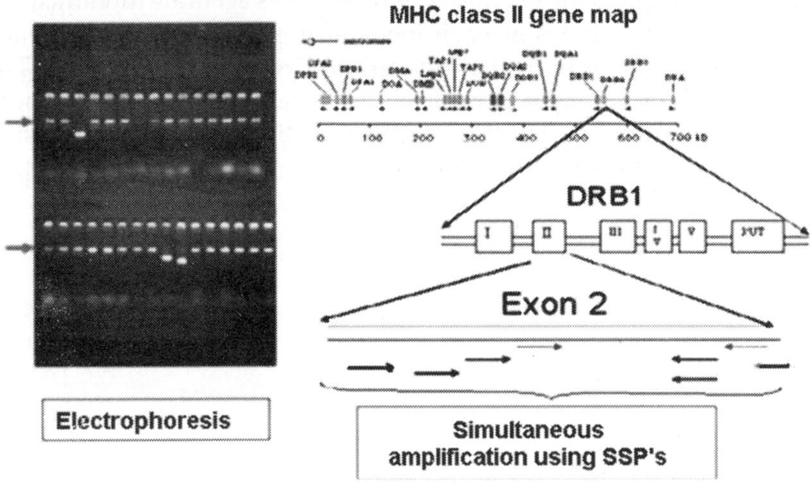

Figure 3.18 Low-resolution PCR with sequence-specific primers (PCR-SSP)

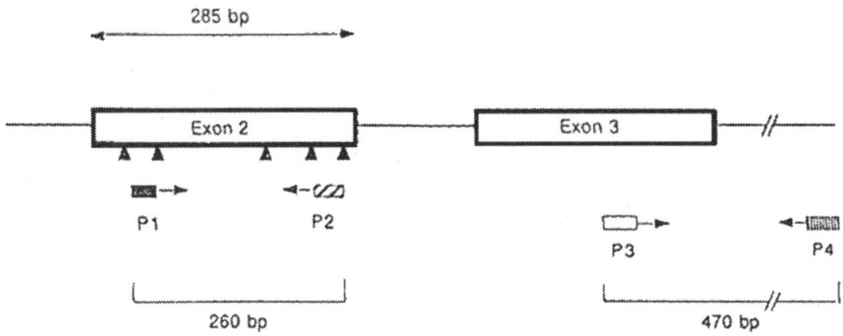

Figure 3.19 Schematic diagram of part of the DRB1 gene that encodes the β chain of

new alleles and therefore data provided by SBT is invaluable. Under this procedure genomic DNA is isolated and purified from peripheral blood.

In the first stage of HLA SBT, HLA locus specific primers are used in PCR to amplify a 2000 bp region of DNA template that represents exon 1 to 5 (exon 1-3 for HLA-DRB1,3,4,5 loci). Each PCR product is then used as template for DNA sequencing of exons 2, 3, and 4 (HLA class I) or exon 2 (HLA-DRB1, 3, 4, 5) using DNA polymerase and individual nested sequencing primers that flank each exon. Sequencing is performed on an automated DNA sequencer using four – color-sequencing chemistry, which allows accurate identification of heterozygous alleles at each locus. Complete sequence for each exon is obtained in both the forward and reverse orientations. The sequence data for each exon are edited, assembled into a linear sequence, and analysed with a variety of additional software to assign the HLA alleles for each HLA locus.

CHAPTER 4

Blood Group Antigens

The ABO system

The original classification of blood into four groups, O, A, B, and AB was based on the presence of blood group substances A or B on red blood cells. Red cells are classified as group A if they are agglutinated by anti-A serum, B if they are agglutinated by anti-B serum, AB if by both anti A and anti B sera and O if by none. The antigens and antibodies of the ABO system are given in **Table 4.1**.

Table 4.1 Antigens and Antibodies in the ABO system

Group	Subgroup	Antigens	Antibodies
O	-	None	Anti-A
			Anti-A1
			Anti-B
A	A1	A+A1	Anti-B
	A2	A	
B	-	B	Anti-A
			Anti-A1
AB	A1B	A+A1+B	Nil
	A2B	A+B	

Inheritance of ABO blood group system:

In 1900 **Epstein** and **Ottenberg** suggested that the ABO blood groups were inherited in a Mendelian fashion and in 1924 **Bernstein** postulated the existence of the allelic genes A, B, O.

Inheritance of ABO blood groups has been established by pedigree analysis. The inheritance of A and B is dominant over O; when both are present genotypically, they are also expressed phenotypically, a pattern known as codominant inheritance. Antigen A is further subdivided into A1 and A2. Mating between A1 and B individuals results in 50% of the gametes inheriting each of the two genes and hence there is equal chance of the occurrence of blood group O, A1B, B and A1 among the offspring's. The ABO genes are located on the long arm of chromosome 9 and linked to the AK1 (adenylate kinase) locus and NP (nail-patella syndrome) locus. The three major alleles are A, B, and O. The A and B genes are responsible for the synthesis of two enzymes, N-acetylgalactosaminyltransferase and B-galactosyltransferase, respectively. The former transfers N-acetylgalactosamine and the latter transfers galactose to preformed carbohydrate structures and confers the A or B antigen specificity. The preformed carbohydrate structure is the H antigen, and inheritance of at least one H gene is necessary for normal ABO group expression. The O gene is believed to be an amorph, producing no detectable product. Therefore, individuals who inherit two O genes express only H antigen on their red cells. Group O red cells demonstrate more H antigen than the red cells of other ABO groups.

Fucose, N-acetylgalactosamine, and galactose are also referred to as the immunodominant sugars for the H, A, and B antigens, respectively. These terminal sugars confer H, A, and B specificity.

The products of the ABO genes are expressed in an autosomal co-dominant fashion. For example, an individual who inherits both A and B genes express both gene products and is of blood group AB. An individual who is having group B could have either inherited two B genes or a B and an O gene. However, family studies are required to know the exact genotype. There is wide variation as genotypes and phenotypes distribution in different racial groups; the B phenotype being more common in blacks than in whites.

Although only three alleles at the ABO locus have been discussed, many more alleles have been described at this locus and have been classified as subgroups of A and B. These often present in the laboratory as aberrant ABO groups. The most common alternate alleles occur for the A gene. It is well established that an A_1 gene and an A_2 gene give rise to group A individuals without aberrant typing. Subgroups for both groups A and B exist.

The ABO genes have been cloned and the A and B cDNA differ by only four amino acid substitutions at positions 176, 235, 266, and 268. Blood group O cDNA is identical to A except for a single base deletion at the N-terminus, resulting in a shift of the reading frame that gives rise to an enzymatically inactive protein product with no transferase activity. Based on this finding, it is postulated that the A gene is the ancestral gene of O.

H Antigen on red cells

The H determinant is the precursor of A and B and is found on all human red cells except those subjects of phenotype Oh. A and B subjects have less H than O subjects.

ABH antigens

Approximately 80% of persons secrete water soluble H and A or H and B antigens in their secretions. Lewis system antigens can also be found in saliva. A second alcohol soluble form of these antigens is present in all tissues of the body except in the brain and on erythrocytes. The secretion of A, B and H antigens in saliva or other body fluids is controlled by a secretor gene (Se) that is inherited independently of the A, B and H genes. Se gene is responsible for secretion status and se for non-secretor status of the individual.

Subgroups of A

Blood group A can be differentiated into two subgroups, A1 and A2 (**Table 4.1**) and 78% of individuals of A blood group belong to subgroup A1. The acceptor site of A1 and A2 gene specified transferase is different. A1 and A2 can be differentiated bio chemically by the lectin *dolichos biflorus,* which reacts with A1 but not with A2.

Classification of the subgroups of A is based on the reactivity of erythrocytes with anti-A, anti-B, anti-H, and anti-A1, the presence or absence of anti-A1 in the serum and the secretion of A and H substance by secretors. Failure to detect a weak A antigen usually results in the erythrocytes being mistyped as group O. Other weak forms of A antigen exist and are of two main groups:

1. Ax (Ax, A4, A5, AO, Ael): Cells of such individuals give negative or weak reaction with anti-A from B donors but good reactions with anti-A from O donors. The serum of such individuals agglutinates A1 cells but not A2. The saliva of secretor individuals of this type contains H substance but usually no A substance.
2. Am : These cells react weakly or not at all with the anti-A from either B individuals or O individuals. The serum lacks antibodies to both A1 and A2. The saliva (if secretor) contains both A and

H. Serologic specificities of various subgroups of A antigen is shown in **Table 4.2.**

Table 4.2 Serologic characteristics of different subgroups of A

Subgroup	Human Anti-A	Anti-A, B	Anti-A_1#	Anti-H	Monoclonal Anti-A	Substance (s) in saliva if secretor	Anti-A in serum
A_1	++++	++++	++++	+w	++++	A and H	No
A_2	++++	++++	0	+++	++++	A and H	4%
A_{int}	++++	++++	++	+++	++++	A and H	No
A_3	++ mixed field	++ mixed field	0	++++	++++	A and H	Few
A_x	0/+w	+/++	0	++++	+++	H	Usually
A_m	+w	+w	0	++++	NT	A and H	No
A_{end}	+w	+w	0	++++	NT	H	No
A_{el}	0	0	0	++++	+	H	Few
A_{bantu}	+w	+w	0	++++	++	H	Usually
A_{finn}	+w	+w	0	++++	NT	H	All

Antibodies of the ABO Blood Group System

Landsteiner observed that most individuals possess antibodies directed against the ABH antigens, which are absent on their own cells. Although these antibodies were named as naturally occurring antibodies, they are formed in response to stimulation by external agents such as bacteria, pollen etc. The immune response to noncellular antigenic stimulation results in the production of the complementary IgM antibodies (Anti-A and / or Anti-B) in immunologically competent individuals and a predictable relationship between antigens and antibodies in the ABO system permits the use of both serum and cell tests in ABO grouping (**Table 4.1**).

ABO blood group antibodies can be of either IgG or IgM type. IgG antibodies are produced as a result of external stimuli as for example following transfusion etc. Antibody levels may vary with age, sex, ethnic origin and season. Various pathological conditions can also alter the antibody titer.

Biochemical Activities related to the development of A, B and H Antigen

The H substance is a mucopolysaccharide, which is found in almost all the tissues and body fluids, like saliva, gastric secretions, bile milk etc. Gene responsible for H antigen expression is the H gene.

The H gene codes for the production of alpha-L-fucosyl transferase, an enzyme that catalyses the addition of L-fucose, the immuno-dominant structure of H antigen, to two slightly differing structures known as the type I and II precursor chains. Once the H gene specified transferase has acted and L-fucose has been added to the chains, the A and B gene specified products then add a monosccharide to the preformed H bearing chains. If a person does not inherit any H gene, L-fucose is not transferred to the precursor substance. As a result, A and B immunodominant sugars cannot be added subsequently. Because the A and B transferase converts the same acceptor molecule, they compete with one another for their substrate. Both A1 and B cells have many immunodominant sugars added, resulting in low level of detectable H antigen on A1 and B erythrocytes.

A gene produces an α - 3 - N- acetyl- D-Galactosaminetransferase that attaches N-acetyl-D-galactosamine to the acceptor molecule. The B gene produces an alpha-3-D galactosy-transferase that attaches a galactosamine to the acceptor molecule. The level of A1-specified transferase enzyme is higher in cord blood than in the sera of A1 (non-pregnant) adults. In pregnancy, the level of A1 specified transferase enzyme in the serum drops to less than half the amount present in the sera of nonpregnant A1 adults.

The transferase produced by the A1, A2 genes are specific for the same monosacharides and use the same acceptor molecule, are both quantitatively and qualitatively different. It has been demonstrated that the level of transferase is 5 to 10 times higher in A1 persons than in A2 individuals and the transferases differ in their chemical requirements such as optimum pH etc.

Most H, A and B antigens synthesized on erythrocyte membrane are attached to glycoproteins, while others are attached to glycolipids. Of all the membrane associated ABO system antigens, about 5% are carried on simple glycosphingolipids, between 10 and 15% on simple polyglycosylceramides, between 10 and 15% on alkali labile glycoconjugates and remaining 65 to 75% on alkali stable glycoproteins. In glycoproteins, N-acetyl-galactosamine joins the oligosaccharide to the peptides; in the glycolipid structures, glucose is the sugar that links the oligosaccharides to the lipid (ceramide) backbone.

Two different precursor chain configurations, known as type I and type II, are found in body fluids and various other tissues. The chains differ only in the linkage of -galactose to a terminal -N acetylglucosamine residue. The carbon 1 of galactose can be attached to either carbon 3 or carbon 4 of -N-acetylglucosamine. The 1-3 linkage represents a type 1 chain while 1-4 linkage represents type 2 chains. The H, A and B immunodominant sugars can be added to either of these chains (**Figure 4.1**).

Extracts made from seeds of *ulex europeans* and *Lotus tetragonlobus* react with RBCs of group O individuals and also with a group specific substance in saliva of O individuals. The degree to which anti H reagent agglutinate red cells of different ABO blood groups follow approximately the order: O> A2 > B> A1.

The Bombay Phenotype

This phenotype arises when two hh genes are inherited at the Hh locus. Such individuals are unable to convert type II paragloboside to H antigen and hence they are unable to make A or B antigens, even though they have inherited two genes at the ABO locus. Being deficient in H, A, and B antigens, these individuals produce anti-H, anti-A, and anti-B as naturally occurring antibodies. On initial testing, Bombay red cells appear to be group O. Their red cells are non-reactive with anti-A, -B, and -AB, whereas they produce anti-H, anti-A and anti-B as naturally occurring antibodies. These individuals cannot be transfused with blood from any donor other than of Bombay group.

Oh is used to identify the Bombay phenotype. If it is possible to determine ABO genes that the individual inherited but is unable to express, the phenotype is written O_h^A, O_h^B, or O_h^{AB}.

Para-Bombay

Individuals who are para-Bombay (like Bombay) are best designated as Bombay non-secretors and Bombay secretors. The red cells of Bombay non-secretors actually have minute amounts of A and/or B antigen expressed on their surface but fail to express any H antigen. These individuals appear to have inherited another allele at the Hh locus, which produces a substrate that serves as an acceptor

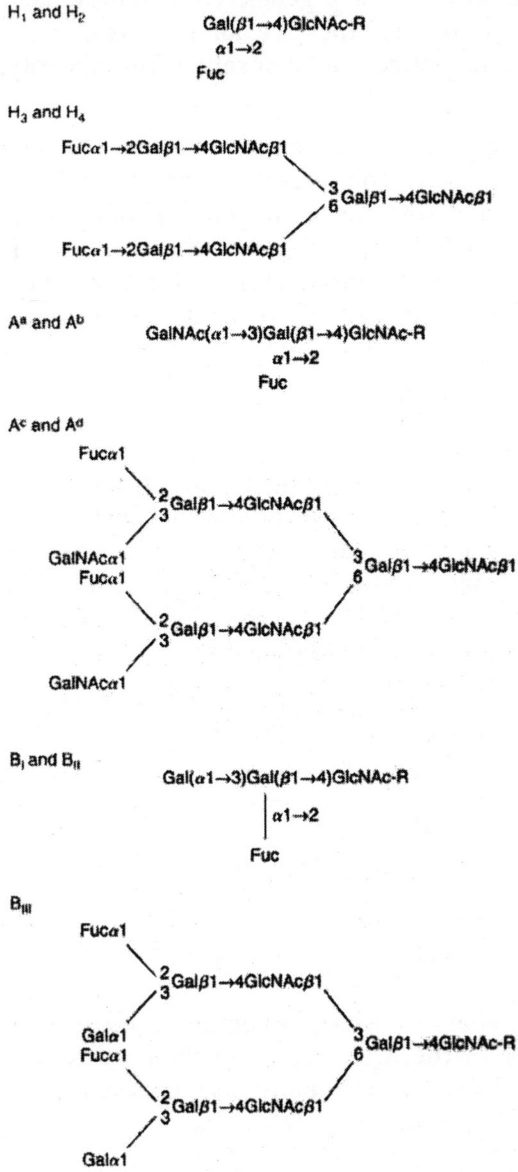

Figure 4.1 Linear and branched ABH antigens. *Only the terminal structures are shown. The number of repeating sugars varies, accounting for the differences in the antigens. Fuc, fucose; Gal, D-galactose; GalNAc, N-acetyl-D-galactosamine; GlcNAc, N-acetyl-D-glucosamine.*

for the products of the A or B genes. These individuals have anti-H in their serum and, because they are non-secretors, they fail to secrete soluble blood group antigens in their saliva. This phenotype is written A_h, B_h, or AB_h.

Inheritance of Bombay secretors may be autosomal recessive or autosomal dominant. The recessive form of Bombay secretor fails to express A, B, or H antigens on the red cell but demonstrate normal amounts of soluble A, B, or H antigens in the secretions. The dominant form of this phenotype demonstrates weak expression of A, B, or H antigens on the red cells and also expresses soluble antigens in the secretions. Anti-H is detectable in the serum of these individuals.

Rh Blood Group System

Rh is the most complex of the blood group systems, embracing over 45 distinct antigens, the absence or presence of which combine to exhibit the Rh blood group type of an individual. The most clinically relevant antigens are D or Rh_o which was discovered in 1940. It has been generally referred to as the Rh antigen, being present in over 85% of the random population. Those individuals that lack the D antigen are considered to be Rh negative.

Two highly homologous and closely-linked genes on the short arm of chromosome 1 encode for Rh antigens. The RHD gene produce the D antigen, or most of its components; and RhCE gene produce the Cc and Ee antigens or their variants. The majority of the antigens within this system represent products of gene cross-over, point mutations or deletions within one or both genes.

The Rh antigens appear to be red cell specific, appearing early during development of red blood cells, and have not been found on other body tissues. Antibodies against the Rh antigens cause severe and fatal transfusion reactions and hemolytic disease of the newborn. The importance of the Rh antigens in the erythroid membrane is exemplified by the fact that in many cases of auto-immune hemolytic anemia, auto-Rh antibodies are frequently found.

The extremely rare (only 32 known throughout the world) individuals who have no detectable Rh antigens are known as Rh_{null} individuals. They have a shortened red cell survival and Rh_{null} cells exhibit stomatocytosis and spherocytosis and have increased permeability to potassium suggesting that they lack a crucial membrane component. A current model suggests that Rh assembles in the membrane as a complex with CD47, LW, RhAG and glycophorin B. Mutations of the RhAG gene accounts for most examples of Rh_{null}

It is ironic that Rh blood group system received the name because it was originally thought to be similar to an antibody produced in rabbits that had been immunized with rhesus monkey cells. By the time it was scientifically proven that they were two distinct antibody specificities, there were too many publications referring to the Rh factor as the product of the D gene and the symbol Rh was too well established to be changed. The antibody produced in rabbits to rhesus monkey cells, and the similar human antibody specificities, have been named after the two original investigators, Landsteiner and Wiener.

Genetics of Rh blood group system

Five major antigens have been found in the Rh system: D, C, c, E and e. C and c, E and e are antithetical while the antithetical antigen for D has never been demonstrated. The term antithetical refers to two antigens controlled by a pair of allelic genes. For convenience the allelic gene to D is expressed as d.

The Rh genes reside on chromosome 1 with the D gene being autosomal dominant. There are two theories regarding the location of these genes. According to the theory of Fisher, there are three closely linked loci inherited as a unit that rarely exhibit crossing over. Each of these loci contains a primary set of codominant allelic genes D and the theoretical d, C and c or E and e. the gene complex of an individual or population may be referred to as the haplotype, which produces an antigen complex with multiple specificities. An alternative proposal of Rh gene inheritance by Wiener remains controversial. He proposed that multiple alleles were positioned at a single complex locus.

Nomenclature of the Rh System

Two types of nomenclature exist for Rh system. These are CDE and Rh-Hr (Table 4.3). The CDE nomenclature of Fisher is used exclusively, but it is some time necessary to use a combination of CDE and Rh-Hr terminology. In the CDE terminology of Fisher the same letter is used for both genes and gene products.

Table 4.3: The Rh gene complex.

Fisher and Race	Wiener
CDE	Rh-Hr
D	Rho
C	rh′
E	rh″
C	hr′
E	Hh
F	Hr
CW	rh^{w1}
G	rh^{G}

The Rh-Hr terminology is based on Wiener's concept that the immediate gene product is a single entity, an agglutinogen, with each agglutinogen characterized by numerous individual serologic specificities called factors. Each factor is recognised by its own specific antibody. What Wiener called agglutinogens are now called haplotypes and what he called factors are now called antigens. Various symbols used are R^1, R^2, R^0, R^Z, r′, r″, r and r^y, which indicate the various alleles. R1 symbolises the presence of C, D and e genes whereas R2 indicates the presence of c, D and E. In most population R^1 (Cde) and r (cde) are common, while r^1 (cde), r″ (cdE) and R^Z (CDE) are rare. R^0 is more common in blacks as compared to whites (Table 4.4).

Table 4.4: Rh Nomenclature

Fisher – Race terms	Genes Rh-Hr Terms	Shorthand Symbols	Frequencies (us) White	Frequencies (us) Black	Gene Product	Antigen Expressed
CDe	Rh^1	R^1	0.42	0.17	R_1	C, D, and e
cDE	Rh^2	R^2	0.14	0.11	R_2	c,D and E
cDe	Rh^0	Ro	0.04	0.44	R_0	c,D and e
CDE	Rh^z	Rz	Rare	rare	R_Z	CD and E
cde	Rh	R	0.37	0.26	r	c and e
Cde	Rh'	r'	0.02	0.02	r'	C and e
cdE	Rh''	r''	<0.01	rare	r''	c and E
CdE	Rh^Y	Ry	Rare	rare	rY	C and E

Rh Antibodies

The Rh antibodies are predominantly of IgG class. IgG1 is the most common type of RH antibody specially it is produced as result of pregnancy. Rh antibodies and their salient features are summarized in **Table 4.5**.

Table 4.5 Rh antibodies and their salient features

Antibody	Mode of reactivity			IgM	Usual immunization sub class IgG	IgA	Usual form of stimulation	
	Sal[1]	AHG[2]	ENZ[3]				RBCs	Others
Anti-D		X	X	Some	Most	Few	X	Rare
Anti-C		X	X		Most		X	Rare
Anti-E	X	X	X	Some	Most		X	Some
Anti-c		X	X		Most		X	
Anti-e		X	X		Most		X	
Anti-C^W	X	X	X		Most		X	Some

1 Sal = Saline
2 AHG = Antihuman globulin
3 Enz = Enzyme

Alterations of Genetic Expression

Expression of inherited genes can be affected by their interactions. When an Rh gene on one chromosome affects the action of another Rh gene on the same chromosome (in terms of increased or decreased antigen production) it is referred to as a cis effect. When an Rh gene on one of the chromosomes of a homologous pair affects the action of an Rh gene on the other homolog, it is called a trans effect.

The cis effect produces a weaker expression of the E antigen in the genotype cDE / cde. The most dramatic Rh transposition effect known to occur is when C is trans to d (Cde/ Cde) which weakens the D antigen.

The Du Phenotype

The red cells of "type D" are recognised by some, but not all, anti-D sera. This is known as D^u. In fact it is considered to be a separate antigen from D. It is less immunogenic than D but can still stimulate an anti-D response in D^u negative persons.

The reactivity of D^u with anti-D is very variable and hence have been classified into high grade D^u and low grade D^u. The high grade

D^u phenotypes may be due to a position effect caused by the presence of Cde or CDE introns on the normal D in the opposite gene complex.

The D Mosaic

The concept of the D mosaic was put forward to account for the fact that some people with D positive erythrocytes produced anti-D that was non-reactive with their own cells. If one or more components of the D antigens are missing from D positive cells, some demonstrate a positive reaction with anti-D sera while others may react weakly with anti - D and be classified as D^u. Phenotypically, erythrocytes of the D mosaic type are similar to low grade D^u and may sometimes be indistinguishable. The distinction, however, is quantitative. This true variant of D is sometimes called D mosaic to distinguish it from other D^u phenotypes.

Another antigen, Go^a, is also part of the Rh system of antigens and is found on erythrocytes that lack part of the D mosaic.

Rh Deletion Phenotypes

There are some rare Rh genes that behave as if portions of the genetic material have been deleted or rendered non functional. The first recognised case was the D-deletion gene d (-). Other gene deletions, E/e and C/c, are generally detected during routine typing.

Rh Null Sydrome

The term **Rh null** is used to describe the phenotype when no Rh antigens are expressed on erythrocytes. If a less complete type of suppressed Rh gene expression is observed, the term Rh mod is used. In both these cases erythrocytes have membrane anomalies and Rh null type individuals have cup shaped red cells (Stomatocyte).

The E Mosaic

The e antigen in some individuals may have a mosaic type structure. The antigen has a high incidence and in most case the relationship of e and anti - e is straightforward. If E+/ e- persons form antibody, it typically reacts with all 'e' positive erythrocytes. Among blacks, however, some persons have red cells that type as e+ but make antibodies that closely resemble anti-e (or anti-f or anti-rh') but do not react with the persons own red cells. The e, hr^s, Hr^B and hr^S -likes (Santiago antibody defined) portions of the e mosaic have been studies and characterized, but do not fully explain the complexity of the e mosaic.

Compound Antigens

Certain combinations of Rh antigens demonstrate a combined effect such as c, e and f antigens and are termed compound antigen. Antibodies against these compound antigens are less frequently found.

Variant Antigens

Although erythrocytes from most individuals with common allo antibodies give straight forward reactions, some produce atypical reactions. Others stimulate production of allo antibodies that do not react with typical Rh antigens. This means they are recognising some subtle difference in composition among various Rh gene products.

The antigen C^W is a common example of variant antigen that usually functions as an allele in the antithetical relationship of C/c. This antigen is expressed in 1% of white Caucasian population. Characteristics of C^W antibody include demonstration of a dosage effect and capability to induce haemolytic transfusion reactions and haemolytic disease of the newborn.

LW Antigen

If rhesus monkey red cells are injected into guinea pigs, the resulting antisera, after partial absorption with human Rh-negative red cells, react strongly with adult Rh-positive red cells and weakly with adult Rh-negative red cells. This antibody was identical to Rh but directed against a different antigen known as LW. Evidence that anti-LW is different from anti-Rh was obtained by Fish and Ford (1942) when they showed that anti Rh sera react with equal strength with Rh-ve and Rh+ve red cells from umbilical cord blood. Further, they found that extracts from either Rh-ve or Rh+ve red blood cells could stimulate the formation of anti-Rh in guinea pigs.

The gene responsible for LW is independent of RH gene; relationship between LW and Rh is still unproved.

The Lewis Blood Group System

In the Lewis blood group system there are two common antigens, Le^a and Le^b controlled by a single gene Le. This system represents a complex interaction between independent Lewis, ABO and secretor genes. The Lewis gene does not code directly for the production of Lewis antigens. Lewis substances are absorbed from the plasma rather than being membrane bound antigens, the Lewis phenotype is influenced by the secretor status of the individual.

There is one functional Le genes, which encodes an alpha - 4 - L - fucosyl transferase that is involved in the synthesis of the Le^a antigen. The Le^b antigenic determinant results from the phenotypic interaction of Le, H and Se genes. The frequency of the Le gene in caucasoid is 0.68 and that of le gene, which has no known gene product is 0.32. The genes are inherited in a classical Mendelian manner.

Mode of Inheritance

The inheritance of the amorph gene, Le in a homozygous state (le le) prevents the production of any Lewis antigen and result in the null phenotype. The Le^b antigen is the product of the interaction of the dominant Lewis gene and the H gene in the presence of the Se gene. This explains why an Le^b gene is not required for the formation of Le^b antigen.

Without Le gene, no lewis substance can be formed and the erythrocytes are Le (a-b-c), regardless of the secretor status.

The results of Lewis gene activity is the production of an alpha-4 fucosyltransferase that catalyses the transfer of fucose to the carbon -4 position of N-acetlylglactosamine to produce the Le^a active structure. The Le^a soluble antigen is then secreted and absorbed on to the erythrocyte membrane.

In the absence of the secretor gene Le^a is not converted to Le^b, therefore, body fluids contain abundant Le^a and no Le^b and the erythrocyte are Le (a+ b-).

If both the H and Le gene products are present, the addition of fucose to the carbon -4 position of N-acetylglucosamine results in a Le^b active structure, because the H gene has transferred fucose to the carbon - 2 position of galactose. This Le^b substance will be present on erthrocytes as well as body fluids and the phenotype is Le (a- b+).

There are two types of carbohydrate chains ending in the glycoprotein of the precursor substance, which are named, chain I and chain II. In type I chains, the terminal galactose has a B-/-3 linkage with N-acetylglucosamine while in type II chains, the sugar has a B-1, 4 linkage. The fucosyltransferase produced by the Le gene is unable to act on a type II chain because the number 4-carbon atom is already occupied by the galactosyl structures. The two antigens, Le^a and Le^b, are derived from type II chains.

The water-soluble Lewis antigens found in secretions are

glycoproteins as are the ABH substances from secretors. These glycoproteins are composed of approximately 80% carbohydrates and 15% amino acids.

Le^a and Le^b antigens are extremely weak at birth or are absent altogether. There are only three major phenotypes of Lewis system. These are Le (a+ b-), Le (a- b+) and Le (a- b-).

Lewis Antibodies

Lewis antibodies are commonly found even in persons who have not received blood or blood component. Anti Le^a Lewis antibodies are sometimes associated with haemolytic transfusion reactions.

The incidence of Lewis antibody in Le (a- b-) persons is 4 times more common in blacks than in whites. Anti Le^a is more common then anti Le^b and is found only in individuals who are Lewis negative. Sera containing anti -Le^a usually agglutinate le (a+) cells suspended in saline and react more strongly at room temperature or colder temperatures. There are two kinds of Le^b, one anti -Le^b antibody that is neutralized by the saliva of all ABH secretors, including those of Le (a- b-) individuals which is termed anti - Le^{bH} and a second form called anti Le^{bL} which is neutralized by the saliva of Le (a- b+) persons.

Biochemical Nature of MN Antigens

M and N are glycoproteins, with approximately 60% carbohydrate, which actively resides on the glycophorin molecule. There are approximately 500,000 copies of the molecule present on each erythrocyte, which are embedded in the erythrocyte membrane. When glycophorin A carries M antigenic activity glycophorin Am, the first amino acid residue is serine and the 5th is glycine. When it carries N activity, glycophorin AV, leucine and glutamic acid replace serine and glycine at positions one and five respectively.

MNS Antibodies

Anti -M and -N antibodies have not been shown to have much clinical relevance. They are usually of IgM type and rarely of IgG type. Anti-S and Anti-s are usually erythrocyte stimulated IgG antibodies. Antibodies of this type can cause haemolytic disease in newborn (HDN) or transfusion reaction.

Inheritance of MN Blood group

The MNSs system is governed by closely linked sets of allelic genes, M, N and S. Because of close linkage between the genes, one of the four possible genetic combinations is inherited from each parent, MS, Ms NS or Ns. This results in 10 possible genotypes. Linkage produces disequilibrium and the gene complex Ns is 5 times more common than NS.

Other Blood Group Systems

Blood Group P

Levine and Landsteiner discovered this at the same time as the MN blood group system. Four antigens are recognized in P system, P1, P2, Pk and p. The mode of inheritance of P blood group system is autosomal dominant. Each of the antigens of the P blood group system has distinctive characteristics. P1 is a glycosphinogolipid on the erythrocyte membrane and is poorly developed on the erythrocyte of new-born infants; it may take years to reach adult levels. Its strength decreases in stored RBCs. Anti-P1 is an IgM antibody with strong non-agglutinating, complement binding activity.

The Luthern Blood Group System

In 1945, an antibody was reported in a patient suffering from a severe dermatological and systemic disorder, now known as lupus erythematosus. This was the Lutheran system which consists of two alleles Lu^a and Lu^b giving phenotypes Lu (a+) and Lu (a-). Antibodies anti Lu^a and Lu^b. Lu (a+) antigen is inherited as autosomal dominant.

The Kell System

It consists of Kell antigens Kp^a and Kp^b and antibodies to these antigens. Kp^a is also known as Penney while Kp^b is known as Rantenberg after the name of the families in which they were discovered for the first time. It is predominantly found in Europeans.

The Duffy Blood Group System

This system was discovered in an individual suffering form haemophilia and had received large number of blood transfusions. The genes recognized are Fy^a and Fy^b. The Duffy alleles Fy^a and Fy^b appear to be inherited as non-dominant autosomal alleles.

Some additional genes have been defined ie. Fy^3, Fy^4 and Fy^5. Antibodies to Duffy blood group are not very common indicating that it is not highly immunogenic.

The Kidd Blood Group System

Discovery and nomenclature of Kidd blood group was the result of haemolytic disease in a newborn child born to Mrs. Kidd Antibodies found in Kidd system are JK^a and anti JK^b. This blood group is most associated with the delayed type of haemolytic transfusion reaction. Most work has been done on anti Jk^a. Kidd blood group system has its importance for population studies because there are striking differences in the frequencies of the phenotype for example the phenotype JK^a has a frequency of 95% in West Africa about 50% among Chinese and 77% among Europeans. The Jk locus has been assigned to the long arm of chromosome 18 (18q11-q12) and there are three alleles at this locus. Very little is known about the biochemical nature of Jk antigens although they are believed to be protein in nature.

Haemolytic disease of the new born

Erythroblastosis fetalis, a hemolytic disease in newborn infants, is caused by blood group incompatibility between mother and child. Although the Rh factor is responsible for the most severe cases of erythroblastosis fetalis, the disease may be produced by any of the other blood group antigens like ABO, Kell, Duffy (Fya), Kidd (Jka and Jkb), and MNSs (M, N, S, and s) systems.

With an Rh-negative mother and an Rh-positive father, the possibility exists that the fetus will be Rh positive. Micro-hemorrhages during gestation permit fetal red blood cells to enter the maternal circulation, causing an immunologic reaction that leads to sensitization of the mother against the Rh factor. Maternal IgG antibodies against fetal red blood cell antigens pass through the placenta into the fetus, where an excessive destruction of fetal red blood cells occurs. When such hemolysis begins during pregnancy, stillbirth may result. While there is little danger of damage to the fetus during the first pregnancy, by the second pregnancy sufficient antibodies will have accumulated in the mother's bloodstream to cause increasing danger of hemolytic disease.

The incidence of erythroblastosis fetalis or hemolytic disease of a newborn varies in different ethnic groups. Rh negativity is higher in whites (15%) than in blacks (5%), and is rare in Asians. The frequency of paternal heterozygosity determines the likelihood of an Rh positive child born to the Rh negative mother. The risk of Rh immunization after delivery of the first child to a nulliparous Rh-negative mother is 16% if the Rh-positive fetus is ABO compatible with its mother, 2% if ABO incompatible, and 2-5% after an abortion. The ABO-incompatible RBCs are rapidly destroyed in the maternal circulation, reducing the likelihood of exposure to the immune system. The degree of Rh sensitization of the mother is directly related to the amount of fetomaternal hemorrhage (ie, 3% with <0.1 ml versus 22% with >0.1 ml).

After sensitization, maternal anti-D antibodies cross the placenta into fetal circulation and attach to Rh antigen on fetal RBCs, these red cells form rosettes on macrophages in the reticuloendothelial system, especially in the spleen. Lysosomal enzymes released by macrophages and natural killer lymphocytes lyse these antibody-coated RBCs; the red cell lysis is independent of the activation of the complement system.

Prolonged hemolysis leads to severe anemia, which stimulates fetal erythropoiesis in liver, spleen, bone marrow, and extramedullary sites, such as skin and placenta. In severe cases, this can lead to displacement and destruction of hepatic parenchyma by erythroid cells, resulting in dysfunction and hypoproteinemia. Destruction of RBCs releases heme, that is converted to unconjugated bilirubin. Hyperbilirubinemia becomes apparent only in the newborn, because the placenta effectively metabolizes bilirubin. HDN due to Kell sensitization results in hemolysis and suppression of erythropoiesis because the Kell antigen is also expressed on the surface of erythroid progenitors.

Hemolysis associated with ABO incompatibility is limited to type O mothers with fetuses who have type A or B blood. In mothers with type A and B blood, naturally occurring antibodies are of IgM class, which do not cross the placenta, whereas in type O mothers, the antibodies are predominantly IgG in nature. Because A and B antigens are widely expressed in a variety of tissues besides RBCs, only a small portion of antibodies that cross the placenta is available to bind to fetal RBCs. In addition, fetal RBCs appear to have less surface expression of A or B antigen, resulting in few reactive sites.

Before the establishment of modern therapy, 1% of all pregnant women in USA developed Rh alloimmunization. Following the introduction of routine prophylaxis for women at risk, incidence of Rh sensitization has declined from 1% of all pregnant women to 11 cases per 10,000 births. Alloimmunization due to Kell antigen accounts for 10% of severely affected foetuses. ABO incompatibility frequently occurs during the first pregnancy and is present in approximately 12% of pregnancies, with evidence of foetal sensitization in 3% of live births. Less than 1% of births are associated with significant hemolysis.

Approximately 50% of infants affected by HDN do not require treatment, while 25% are born close to term but have severe jaundice. These children may either die (90%) or become severely affected by kernicterus (10%) if they are not treated. The remaining 25% of affected fetuses develop hydrops in vitro. About half of newborns are affected before 34 weeks of gestation, and the other half are affected between 34 weeks of gestation and term. Before any interventions were available, the perinatal mortality due HDN was 50%. Exchange transfusion was introduced in 1945 and it reduced the perinatal mortality rate to 25%. Later, concept of early delivery was introduced for those of severely affected non hydropic fetuses by 34 weeks of gestation followed by prompt exchange transfusion. This decreased mortality rate to 16%. Today all women are investigated in the first pregnancy and those at risk for allo immunization undergo an indirect Coombs test and antibody titers during their first prenatal visit. If the Coombs test is positive, the paternal blood group genotype is established and serial maternal titers should be established if the father is homozygous.

CHAPTER

5

Immunoglobulin Structure and Function

Antibodies are antigen specific products of B cells and collectively form a family of plasma proteins known as **immunoglobulin.** They are mediators of the humoral immune response and bear the essential features of adaptive immunity, that is, **diversity** and **specificity**. Antibody production is the defence mechanism against invading foreign substances and the body has evolved a variety of mechanisms, each dependent on a somewhat different property or function of an immunoglobulin molecule. Thus, when a specific antibody molecule combines with a specific antigen or a pathogen, several different effects or mechanisms come into play. These differences are due to the different classes of immunoglobulins (**isotypes**), each of which may combine with the same epitope, to trigger a different response.

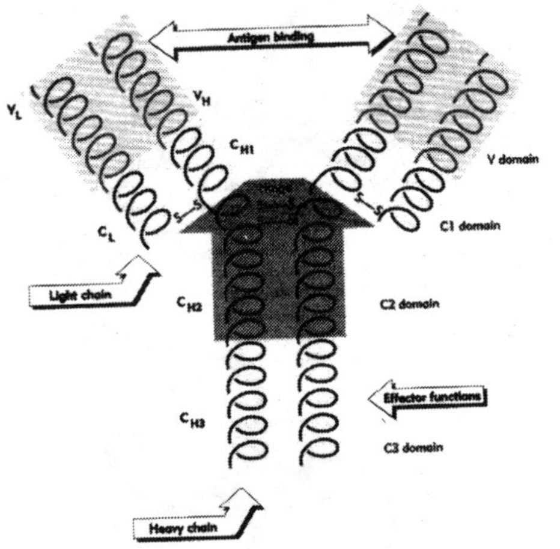

Figure 5.1 Schematic diagram of an immunoglobulin molecule

Each immunoglobulin (Ig) molecule is made of 4 polypeptide chains, two heavy chains and two light chains. The heavy are longer than the light chains. These chains are held together by noncovalent forces and covalent interchain disulphide bridges that form a bilaterally symmetrical structure (**Figure 5.1**). All immunoglobulins conform to this basic structure with additional structures added to some for specialised functions. The heavy (H) and light (L) chains are composed of a number of folded globulin domains. Each domain is 100-110 amino acids in length and contains one intra-chain disulphide bond. Light chains contain two such domains while heavy chains contain 4 or 5 domains.

Amino acid sequences of different Ig molecules have shown that heavy and light chains of different immunoglobulins vary widely. However, this variability is confined largely to the N-terminal domains, whereas sequences of other domains remain relatively constant. Therefore, the N-terminal domain of the heavy or light chain is termed as variable region or VH and VL. The other domains are together called constant region or CH and CL. The light chain is designated as kappa (*k*) or lambda (l). No functional differences have been found between antibodies utilizing kappa or lambda chains. However, it is interesting to note that the ratio of antibodies utilizing kappa or lambda light chains varies between different species.

Within an immunoglobulin unit, the heavy and light chains are aligned in parallel (**Figure 5.2**) with the VH and VL domains and CH1 and CL domains lying adjacent to one another.

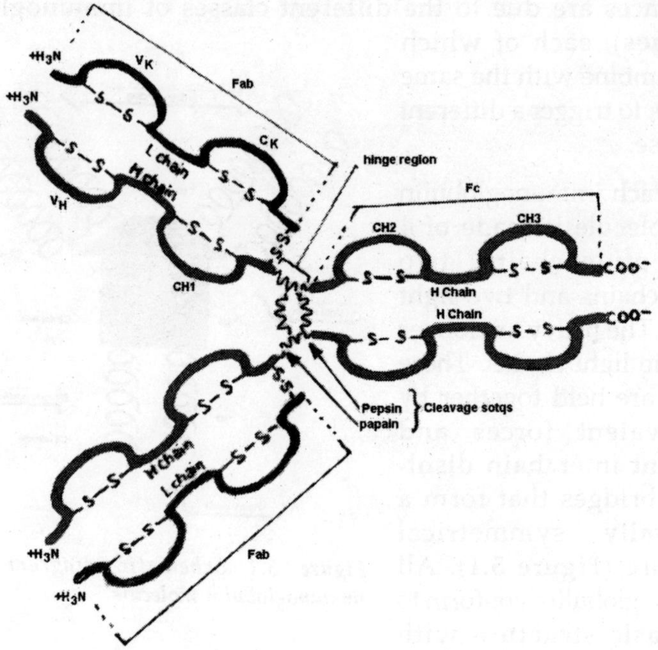

Figure 5.2 Schematic diagram of an IgG (k) human antibody molecule showing the basic 4- chain structure and domains.

One or more interchain disulphide bonds link the heavy and light chains. The remaining CH domains of the two corresponding heavy chains align alongside and may be linked by disulphide bonds. This gives the molecule a T or Y-shaped appearance which can be appreciated in electron micrographs of the protein. The region at the base of each arm of the T or Y, located between CH1 and CH2 domains is called the **hinge region**. It has been so termed because it provides a certain amount of flexibility to the molecules.

Proteolytic enzymes, papain & pepsin have helped to understand the structure and function of these molecules (**Figure 5.3**). The hinge region of the molecule is most susceptible to cleavage by these enzymes. Papain cleaves the molecule on the N-terminal side of the inter-chain disulfide bonds. This breaks the molecule into 3 fragments of almost equal size. Two of them are identical and retain antigen-binding activity and are termed '**Fab**' fragments for 'fragment antigen binding'. Each of them consists of one entire light chain and the VH and CH1 domains of the heavy chain. The third fragment has no antigen binding activity but crystallises easily; hence it is termed the **'Fc' fragment** or **fragments crystallisable**. This corresponds to the paired CH domains of the heavy chain other than CH_1. Pepsin, cleaves

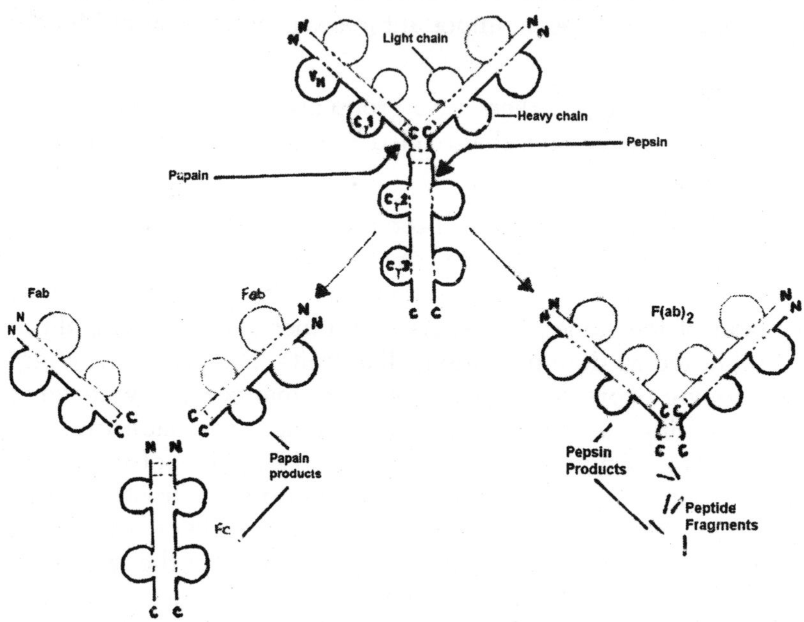

Figure 5.3 Proteolytic fragments of an IgG molecule

the Ig molecules on the carboxyl terminal side of the disulphide bridges. This yields a single large fragment, F $(ab)_2$, which contains both the light chains and VH and CH1 chains linked by the disulphide bonds. The Fc region is extensively degraded by pepsin into multiple fragments.

Differences between the different Ig isotypes result from structural variations in heavy chains, which have generated domains that mediate a variety of functions. A genetic variation among individuals in the structure of immunoglobulin is termed allotype. It depends on the existence of allelic forms (allotypes) of the same protein, as a result of the presence of different forms of the same gene at a given locus. As a result of allotypy, a heavy or light chain constituent of any immunoglobulin can be present in some members of a species and absent in others. This situation contrasts with that of immunoglobulin classes or subclasses, which are present in all members of a species.

Allotypic differences at known loci usually involve changes in only one or two amino acids in the constant region of a chain. With a few exceptions, the presence of allotypic differences in two identical immunoglobulin molecules does not generally affect binding with antigen, but serves as an important marker for analysis of Mendelian inheritance.

The allotype markers are a group on the γ chain of human IgG called Gm, a group on κ chain called Km, and a group on the α chain, called Am.

Allotypic markers have been found in the immunoglobulins of several species, usually by the use of antisera generated by immunization of one member of a species with antibody from another member of the same species. As with other allelic system, allotypes are inherited as dominant Mendelian traits. The genes encoding the markers are expressed in a co-dominant manner. As we have seen, the combining site of a specific antibody molecule is made up of unique combination of amino acids in the variable regions of the light and heavy chains. Since this combination is not present in other antibody molecules, it should be immunogenic and capable of stimulating an immunologic response against itself in an animal of the same species. Oudin and Kunkel, in early 1960s, showed that experimental immunization with a particular antibody or myeloma protein could produce an antiserum specific only for the antibody that was used to induce the response and for no other immunoglobulin of the species. These antisera contain populations of antibodies specific for several epitopes, called, idiotopes which are present in the variable (heavy and light) region of the antibody used for inoculation. The collection of all idiotopes on the inoculated antibody molecule is called the idiotype (Id). In some cases, anti-idiotypic sera prevent binding of the antibody with its antigen, in which event the idiotypic determinant is considered to be in or very near to the combining site itself.

The Light Chains

There are two types of light chains, Kappa (*k*) and Lambda (λ), which occur with a 2:1 ratio in humans and 20:1 ratio in mice. A light chain has a molecular weight of approximately 23,000 daltons and the *k* and λ chains differ in the CL region sequences. They have identical functions and either may associate with any one of the heavy chain classes to form a complete Ig molecule. A given immunoglobulin molecule always contains exclusively *k* or λ chains, never both. Also, it must be emphasised that all the progeny of one B cell will produce only one type of light chain. However, the light chains produced by different B cell differ in their VL region, giving each antibody its antigen specificity.

The Heavy Chains

There are 5 different classes (or 150 types) of heavy chains. These differ in their CH region sequence as well as their physical and biological properties. However, all the heavy chains in any given immunoglobulin molecule are identical and different immunoglobulin vary in the CL region as is seen in the light chains. The five classes of heavy -chains are mu (μ), delta (δ), gamma (γ), alpha (α), epsilon (ε) and the immunoglobulin which contain these heavy chains are designated IgM, IgD, IgG, IgA and IgE respectively. The γ chain may be further divided into 4 subclasses (γ1,γ2, γ3, and γ4) and the α chain into 2 subclasses (α1 and α2) on the basis of minor differences in the CH sequences.

The heavy chain polypeptides range in molecular weight from about 50,000 to 70,000.

Hinge Region

The gamma (γ), delta (δ) and alpha (α) heavy chains contain an extended peptide sequences between the CH1 and CH2 domains, that has no homology with the other domains. This region, known as the **hinge region**, provides IgG, IgD and IgA with segmental flexibility. As a result, the two Fab arms can assume various angles relative to each other. When an antigen is bound, the angle between the arms of the Y shaped antibody molecule varies in different complexes, reflecting the flexibility of the hinge region. This allows the 'Fab' arms to move and twist to align the antigen combining region with epitopes displayed on antigen surface, and the Fc to move and maximize various effects or functions such as complement activation or binding to Fc receptors.

Two prominent amino acids in the hinge region are proline and cysteine. Proline provides extended polypeptide conformation, making the hinge region particularly vulnerable to cleavage by proteolytic enzymes. The disulphide bonds vary in numbers and location within the hinge region from one type of immunoglobulin to other and also between species.

Membrane bound and Secreted Immunoglobulin

Immunoglobulin of all classes can exist as either membrane bound form on B cells or in secreted form as soluble Ig. Membrane immunoglobulins always exist as individual 4-chain units and are heavier in molecular weight compared to secretary forms as they contain additional amino acids residues. There are 26 hydrophobic amino acids that span the lipid bilayer and the transmembrane regions of membrane Ig heavy chains show a high degree of homology. They all adopt the conformation of a helix as they span the lipid bilayer. The hydrophobic cytoplasmic tail varies in length depending on the class of the heavy chain and is composed of three amino acids. λ chain has 28 amino acid residues in its cytoplasmic tail.

The secretary forms of μ, α and δ heavy chains, but not of γ and ε chains have additional extended non globular sequences. The carboxyl terminal side of the last CH domains is called **tail piece**. In secreted IgM and IgA molecules, they contribute to intermolecular interactions that result in multimeric molecules. Less is known about circulating IgD since it is present only in trace amounts in blood.

Three-Dimensional Structure of Antibody Molecule

In 1969, Edelmen and Gall proposed that each homology region is folded into a compact globular structure and linked to neighbouring domains by more loosely folded sections of peptide chain. It was also suggested that each domain had evolved to fulfil a specific function such as antigen binding (VL+VH domains) or complement fixation (CH2 domain).

All immunoglobulins have the same basic folding pattern. It is like a cylindrical sandwich consisting of one layer of four strands of peptide chains and other of three strands of peptide chains. In both layers, adjacent segments run in opposite directions and the two layers are held together by the intra-chain disulphide bridges and hydrogen bonds in β pleated sheet configuration. The amino acid side chains sandwiched between the layers are largely of a hydrophobic nature in all the domains; the amino acid side chains facing outwards from the constant domains differ from those of the variable domain.

In case of the variable region domains, heavy and light chain components (VH and VL) interact with each other differently. The four strand layers are now facing outwards and the three-strand layers interact with each other across the water filled region, which becomes the antigen-binding site. In areas where the structural integrity of the binding site requires it, hydrophobic residues appear to have displaced the three strand sections.

The Antigen Combining Site

The antigen-combining site is composed of loops of the hypervariable regions, which come together to form a continuous surface, interacting with the antigen through a combination of hydrogen bonding, electrostatic interactions and Van-der Waals forces. The hypervariable site is at the tip of the Fab segment. Since these regions together form the binding site and determine specificity by forming a surface complementary to the antigen, they are also called **complementarity-determining regions (CDRs)**. The antigen antibody binding has been studied in some detail in the case of a few antibody-hapten systems. For example, the mouse myeloma protein, MOPC 603, has a binding affinity for phophorylcholine, and X-ray diffraction analysis has been used to establish the nature of the forces involved in this interaction. The hapten is much smaller than most complex antigenic determinants and so does not fill the entire binding cleft of the antibody molecule. It is off set towards the hypervariable region of the heavy chain and most of the interactions that have been mapped are with heavy chain residues. The negatively charged phosphate group of the hapten interacts with the positively charged NH2 group of an arginine residue at position 52 of the heavy chain, forming a strong hydrogen bond with an OH group of a neighbouring tyrosine residue (heavy chain residue number 33). The choline group has a positive charge and is close to two negatively charged glutamic acid residues (one at position 35 in one hyper variable region of the heavy chain and the other, at position 58, in the second hypervariable region). In addition, there are Van-der Waal's interactions between the choline group and carbon atoms in both heavy and light chains.

Immunoglobulin Isotypes

The five major classes of immunoglobulins differ from one another in the amino acid sequences of the heavy chain constant regions. This results in structural and functional differences among different isotypes. The physical properties and biological activities of the immunoglobulin isotypes are listed in **Tables 5.1**.

Immunoglobulin G (IgG)

IgG is the major immunoglobulin in normal serum. In humans, it accounts for 70-75% of the total immunoglobulin pool and is the most abundant antibody produced during a secondary humoral immune response. It is a monomeric protein with a molecular weight of 146,000 daltons and a heterogeneous electrophoretic mobility. It consists of two γ chains and two light chains. The first human heavy chain to be sequenced completely was γ chain of a mouse myeloma protein. The CH2 and CH3 domains of the four human γ chain subclasses have been sequenced and show a remarkably high degree of sequence homology ranging from 92 to 98%. Each γ chain consists of 4 domains (VH, CH1, CH2 and CH3) of which VH and CH1 associate with Ck / Cλ and CH2 or CH3 of the 2 heavy chains associates with one another. Adjoining the heavy chain, the molecule has one oligosaccharide chain attached to the CH region. There are four subclasses of IgG (**Table 5.2**) and the structural differences between the subclasses are in the hinge region between the CH1 and CH2 domains. These differences may be in the size of hinge region and number and positions of the interchain disulfide bonds between the heavy chains; there are two of them in IgG1 and IgG4 and four in IgG2 and IgG3 (**Figure 5.4**). IgG1 and IgG3 easily cross the placenta. IgG3 is the most active immunoglobulin involved in complement activation while IgG1 and IgG2 are relatively inefficient at complement activation; IgG4 is not able to activate complement at all. IgG1 and IgG3 binds to Fc receptors on phagocytic cells with a high affinity. They bind through the CH3 domain to the Fc receptors.

Table 5.1: Physical, chemical and biological properties of human Ig class

Molecular form	**IgM**	**IgD**	**IgG**	**IgA**	**IgE**
	Pentamer hexamer	**Monomer**	**Monomer**	**Monomer**	**Monomer**
No. of C region domain	4	3	3	3	4
Tailpiece	+	-	-	+	-
Accessory chains	J Chain SC	None	None	J chain, SC	None
Subclasses	None	None	G_1, G_2, G_3, G_4	A_1, A_2	None
Molecular weight	950 kD 1150 kD	175 kD	150 kD	160 kD 400 kD	190 kD
Carbohydrate content (%)	10	9	3	7	13
Percentage of hold serum IgE	5 – 10	0.3	75 – 85	7 -15	0.02
Average adult free serum level (mg/ml)	0.7 – 1.7	0.04	95 – 125	1.5 – 2.6	0.0003
Placental transfer	-	-	+	-	-
Classical Complement activation	++	-	+	-	-
Alternate complement activation	-	+	+	A_1+, A_2-	-

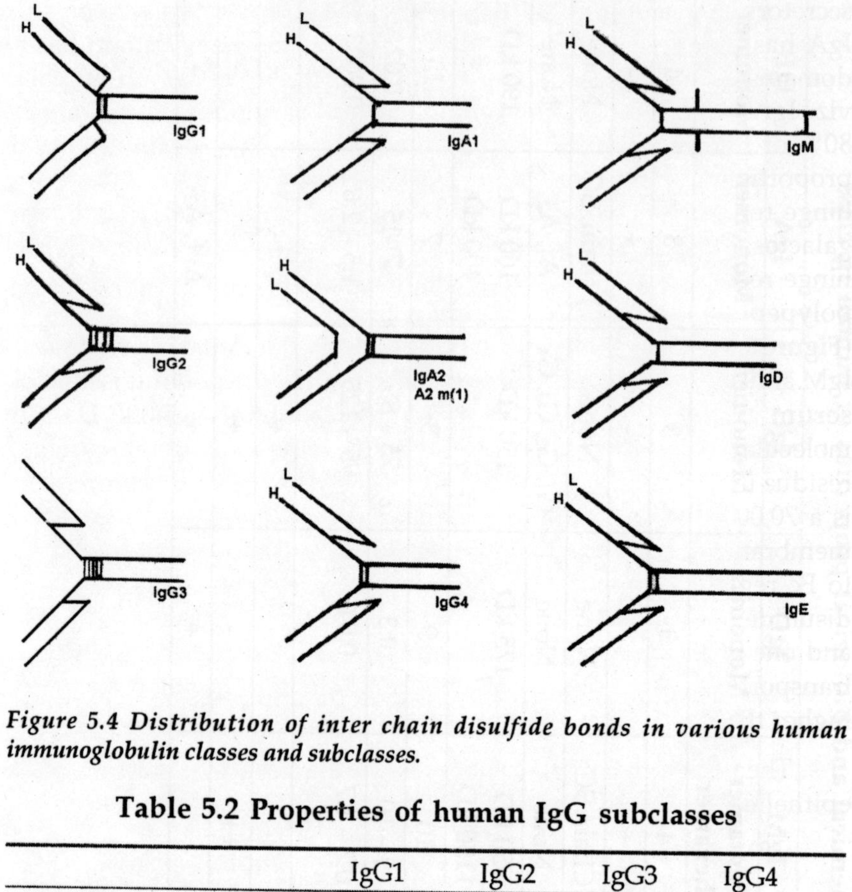

Figure 5.4 Distribution of inter chain disulfide bonds in various human immunoglobulin classes and subclasses.

Table 5.2 Properties of human IgG subclasses

	IgG1	IgG2	IgG3	IgG4
Abundance (% of total IgG)	**70**	**20**	**6**	**4**
Half-life in serum (days)	**23**	**23**	**7**	**23**
Placental passage	**+++**	**+**	**+++**	**+++**
Complement fixation	**+**	**+**	**+++**	**-**
Binding to Fc receptors	**+++**	**+**	**+++**	**-**

Immunoglobulin A

In humans, IgA accounts for about 15-20% of the total serum immunoglobulin pool. It is the predominant class in human secretions and occurs in breast milk, saliva, tears and mucosal surfaces. IgA primarily occurs as monomers, but are also seen as dimers, trimers and even tetramers. IgA present in the external secretions, known as secretory IgA usually occur in dimeric or tetrameric forms. Monomeric IgA has a molecular weight of 160,000 and the heavy chain has 4 domains as in γ chains. There are two subclasses of IgA in humans viz. IgA1 and IgA2. IgA1 predominates in serum and forms about 80% of total IgA. In seromucosal secretions they occur in equal proportion. Differences between IgA1 and IgA2 are present in the hinge region. The α -2 chains lack 12 amino acids residues and two galactosamine rich oligosaccharide units, which are present in the hinge region of α-1. The IgA molecule also has an attached **J chain** polypeptide, and a polypeptide chain called secretory component (**Figure 5.5**). J chain polypeptide is identical to that found in pentameric IgM and serves a similar function in facilitating the polymerisation of serum and secretory IgA. It is of approximately 15,000 Dalton molecular weight and is disulphide bonded to the penultimate cysteine residue in the tail piece of the α or μ chains. The **secretory component** is a 70,000 MW polypeptide produced by epithelial cells of mucous membranes. It consists of five immunoglobulin like domains that bind to Fc regions of the IgA dimer. This interaction is stabilised by a disulfide bond between the fifth domain of the secretory component and one of the chains of the dimeric IgA. It is presumed to facilitate transport of IgA through epithelial cells. Daily production of IgA is higher than any other immunoglobulin class.

The **poly-Ig receptor** is expressed on the surface of most mucosal epithelia, (digestive, respiratory and genital tracts) and glandular

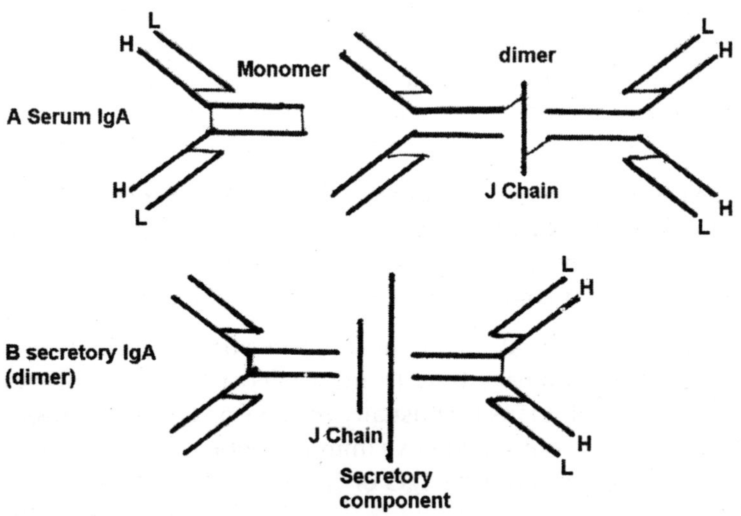

Figure 5.5 Schematic diagram of secretory and serum IgA

epithelia in mammary, salivary and lacrymal glands. IgA polymers bind to the receptor; the receptor-IgA complex is transported across the epithelial cell to the luminal surface following which the poly-IgA receptor is enzymaticlly degraded, releasing the secretory component bound to dimeric IgA.

Immunoglobulin M

IgM occurs predominantly as a pentameter, $(H2\ L2)_5$, with a molecular weight of 970,000 daltons. Significant amounts of a low molecular weight monomeric form occur in the serum of some patients with immunological disorders, in particular Waldenstrom's macroglobulinaemia and ataxia telangectasia.

Structurally, IgM consists of five subunit with the heavy (μ) and light chains polymerised in association with one J chain. Each μ chain has five inter chain domains. The μ -chain is rich in carbohydrates, having five oligosaccharide groups per chain. Each is attached to an asparagine residue in the obligatory sequence, Asn-X-Ser/Thr and both simple and complex side chains occur. They may be responsible for increasing the solubility of the molecule and to influence its conformation.

IgM is the first immunoglobulin class produced during a primary response to an antigen and is also the first immunoglobulin to be synthesized by neonates. Because of its pentameric structure with 10 antigen-binding sites, serum IgM has a higher valency than the other isotypes. An IgM molecule can bind to small hapten molecules. However, because of steric hindrance, only 5 molecules of larger antigens can be bound simultaneously. The increased valency of pentameric IgM increases its capacity to bind such multidimensional antigens as viral particles and red blood cells.

IgM is the most common immunoglobulin expressed on the surface of B-lymphocytes. It is the Ig class most frequently found as "natural" antibodies for example against blood group antigens, and is the most efficient complement fixing immunoglobulin.

IgM is found in all vertebrates but its structure differs as one move up the evolutionary tree. In most teleost fish the molecule is tetrameric, in amphibians it is usually hexameric and in reptiles and mammals it is pentameric. Primitive immunoglobulins are more related to human IgM than to human IgG or IgA.

Immunoglobulin D (IgD)

It forms less than 1% of the serum immunoglobulin pool. Its molecular weight is 184,000 daltons. IgD is a monomeric immunoglobulin with a single disulphide bridge between the two chains. It is a labile protein and is likely to undergo spontaneous proteolysis by plasmin if 2- amino carproic acid is not added as an enzyme inhibitor during isolation. From sequence homologies with other classes, CH1 and CH3 domains are idiotypes. These regions are separated by a long hinge segment, which lacks the cysteine bridge found in human α chains. There is, however, a disulfide bridge between the chains in the carboxy terminal tailpiece of the secreted molecule. In cells that co express IgD and IgM, both classes of heavy chains are produced by alternate splicing of a single RNA and have identical antigen specificity. The physiological function of IgD class of immunoglobulin is still not very well understood.

Immunoglobulin E (IgE)

It constitutes a very minor component of serum immunoglobulins (0.004%). It is a monomeric molecule with a molecular weight of 188,000 daltons and a five-domain heavy chain structure. It has no hinge region and hence is least flexible of all the immunoglobulins (**Figure 5.4**). No allotypes are known. Carbohydrate units are attached to the heavy chain throughout and the carbohydrate content is 13%. Serum concentration of IgE is only 0.03 mg/100ml.

It is the major class of antibody involved in arming mast cells and basophils for their role in allergic reaction. IgE interacts through its Fc region with Fc receptors on mast cells; cross-linking by antigen activates signallingwhich results in release of the mediators of allergy like histamine. CD23 plays a very major role in IgE mediated cytotoxic reaction against parasites.

IgE, because of its higher molecular weight is not able to cross the placenta. It plays a major role in parasitic infections and its serum levels are raised during a large majority of parasitic diseases. Immune complexes containing IgE do not activate complement.

Biosynthesis of Immunoglobulin

Biosynthesis of immunoglobulin chains occurs on polyribosomes, sedimenting at about 200S, whereas other H chains are synthesized on larger polyribosomes sedimenting at 300S. Various intermediate forms are generated during the assembly of the four chain molecule and these play a role in the control of synthesis. Freshly synthesized light chains are released from polyribosomes into the rough endoplasmic reticulum. Both light and heavy chains insert into the rough endoplasmic reticulum via their hydrophobic leader sequence. Leader sequences situated on the N terminal end of the growing polypeptide chain are later on cleaved and form the completed heavy and light chains which are then released from the internal space of RER and assemble into four chain structure. Heavy chains lacking the transmembrane region are transported as soluble molecules in the vesicles. Time taken for secretion of Ig molecule is about 30 minutes and mature plasma cells can produce about 10^5 antibody molecule per minute.

The combining site of an antibody has a complementary stereo chemical structure to a particular portion of the antigen, which is called the epitope (or antigenic determinant). Because of their close fit, antibodies and antigens form stable complexes, even though they are joined only by covalent forces that individually are quite weak.

An antigen typically contains a number of different epitopes and, thus, stimulates the production of a variety of different antibodies. Even if an antigen contains only a single epitcpe, its introduction into the body would still generate numerous antibody species, ali capable of binding to the antigen, but in different configurations having different affinities. Because individual antigens can interact with different B-cell clones containing different membrane-bound immunoglobulin, the response is said to be polyclonal. A single species of antibody, called a monoclonal antibody, directed against a specific epitope can be prepared *in vitro*. Monoclonal antibodies have proven invaluable tools in basic and applied biomedical research and are now being widely used as therapeutic molecules.

The hypervariable portions of light and heavy chains determine the antibody combining site and specificity whereas the remaining portions of the variable domains provide a scaffold that maintains the overall structure of the combining site. The differences in length and sequence between different Ig isotypes enable antibodies of various classes to carry out different biological functions. For example, heavy chains of an IgM molecule bind and activate one of the proteins of the complement system, which leads to the lysis of bacterial cells to which the IgM molecules are bound. Heavy chains of IgE molecules play an important role in allergic reactions by binding to specific receptors on the surfaces of mast cells, triggering the release of histamine. In contrast, heavy chains of an IgG molecule bind specifically to the surface receptors of macrophages and neutrophils, inducing these phagocytic cells to ingest the particle to which the antibodies are bound. The heavy chains of IgG molecules are also important in allowing this class of antibodies to pass from the blood vessels of a mother to those of her foetus during pregnancy. While this provides the foetus and newborn with passive immunity to infectious organisms, it may cause a life-threatening erythroblastosis fetalis if the IgG is anti-D against blood group antigen D.

Evolution of Immunoglobulins

Immunoglobulin molecules emerged during evolution approximately 450 million years ago and are found from hagfish and lamprey to vertebrates. Higher cartilaginous fishes, which arose 350 million years ago, have both monomeric immunoglobulin and high molecular weight IgM like molecules. In the shark fish pentameric IgM, but not the monomeric IgG, incorporates a J chain like molecule. More significantly, sequencing of the variable region of shark light and heavy chains, reveal that these are more closely related to their mammalian equivalents than they are to each other. Therefore, it appears that the VL and VH domains diverged more than 400 million years ago. There is marked difference between amphibian immunoglobulin and mammalian IgG.

There is also evidence of a low molecular weight immunoglobulin (called IgN) in reptiles and birds; but there are large species variations in its occurrence. It is a major immunoglobulin in duck but minor in chicken.

A molecule resembling IgA has been described in birds and most mammals. Similarly, an IgD like molecule has been demonstrated in birds and mammals and suggested in reptiles. These observations may be suggestive of evolutionary origin of IgD before the divergence of birds and mammals from reptiles more than 180 million years ago. There is evidence for the existence of an avian IgG class but further work is required in this area.

Minimum mutational distances have been calculated from known amino acid sequences. These calculations permit estimates of relatedness of same protein in different species and possible relationships between apparently diverse proteins. All immunoglobulins have evolved from a primitive unit of 110 amino acids by process of gene duplication.

CHAPTER 6

Immunoglobulin Genes

The immune response has great diversity, requiring many different antibody molecules, each with a unique affinity and specificity. The immunoglobulin genes have evolved to cater to this diversity and meet all variability of antigens. Two theories were put forward to explain the genetic basis of antibody diversity. The germ line theory postulated the existence of different genes to code for each antibody light or heavy chain. This theory was difficult to reconcile with the vast number of possible antibody molecules an animal could make; far too much of the genome would have to be devoted to the immune system. The **'Somatic Theory'** postulated a smaller number of light and heavy chain genes that would be inherited in the germ line. In this model, most of the diversity would be generated by mutations causing an expanding repertoire of novel antigen binding specificities.

Dreyer and Bennett proposed in 1965 that two separate pools of genes, one for variable region and other for constant region encoded each immunoglobulin polypeptide. Products of different variable region genes could combine with the same constant region gene, generating the repertoire. It was postulated that variable and constant region genes joined together before they could be transcribed and translated into protein. In humans, immunoglobulin (Ig) heavy chain genes are located on chromosome 14 at position 14q 32 and light chain genes for k and λ at 2p12 and on 22q11 respectively. The murine immunoglobulin genes are found on chromosome 12 for heavy chain and on 6 and 16 for light chains *k* and λ respectively. Direct experimental evidence was provided by Susumi Tonegawa, who showed that antibody gene are assembled by recombination events that move DNA encoding for variable regions closer to DNA encoding the constant regions. DNA isolated from embryonic mouse cells, and adult cells, were digested with a restriction endonuclease, electrophoretically separated and hybridized with a nucleic acid probe for the Ig gene isolated from a myeloma. The embryonic DNA showed two different size fragments while adult DNA, revealed only a single fragment, clearly indicating that separate gene segments in the embryonic cell had rearranged to form the functional transcript in the differentiated cell (**Figure 6.1**). These pioneering experiments of Tonegawa, for the first time demonstrated the phenomenon of recombination and fetched him the Nobel Prize in 1987.

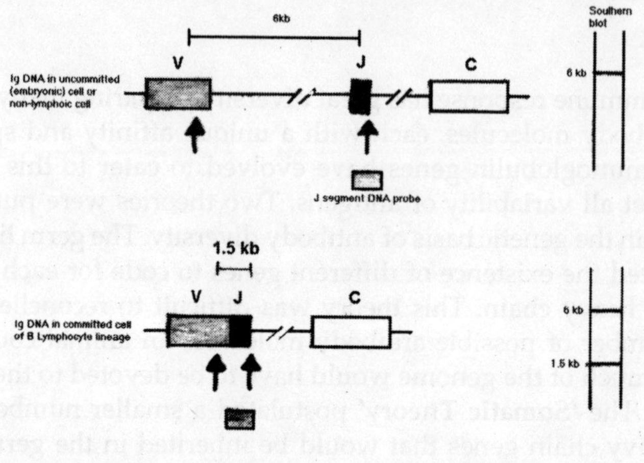

Figure 6.1 Detection of Ig gene rearrangements by southern blot hybridisation

The different genes encoding the two light chains and the heavy chain genes have a similar basic organisation (**Figure 6.2**). Each locus has multiple genes that give rise to the V and C regions of the Ig molecules. At the 5' end of each locus is the V gene exon, each of which is about 300 base pairs (bp) in length.

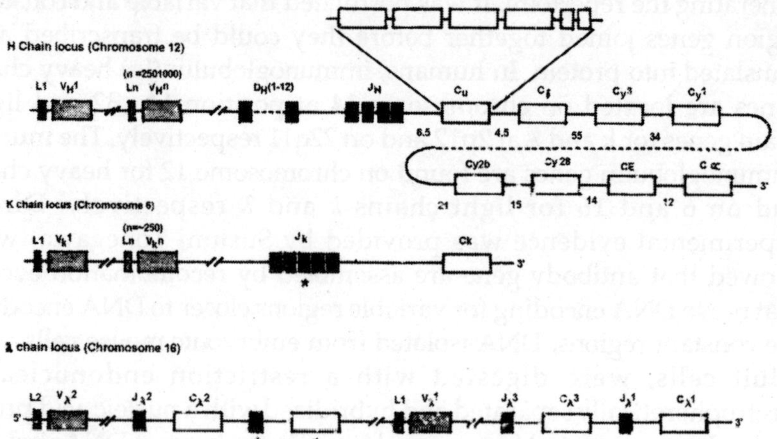

Figure 6.2 Organization of mouse Ig genes in the germ line

They are separated from one another by non-coding stretches of DNA. Approximately 90 bp to the 5' of each V region exon is a small 60-90 bp long sequence that codes for the translation initiation signal and 20 to 30 amino terminal residues of the translated protein. These form the signal (**or leader**) peptide, which guides the newly formed polypeptide during their synthesis on ribosomes into the lumen of the ER. Within the ER, the signal sequences are rapidly cleaved and lost.

The number of V genes may vary from 2 in the murine λ chains to 1000 in murine H chains. Human H and L chains have 100 - 200 V genes (**Table 6.1**). The V genes are separated by varying distances on their 3' end from the C region.

Each H chain constant region actually consists of 3 to 4 exons that code for carboxy terminal end including the transmembrane (TM) and cytoplasmic domains. Lying between the V and C genes are additional coding sequences, the joining segment (J) and the diversity (D) segment in H chains. The D and J genes code for the carboxy terminal ends of the V regions including the third hypervarible region of the antibody molecules. Thus the *k* and λ chain proteins are encoded by V and J exon for the V region along with the C exon while the H chain protein is encoded by V, D and J exons for the V region along with C exon. There is a dose relationship between exon intron structure in the DNA and the domain structure of the protein.

Table 6.1 Gene segments in the Ig regions of mouse

Gene Segment	No. of gene segments	
	Human	Mouse
V_H	~ 100	> 300
V_K	> 70	90 -300
V_λ	20 - 30	2
D_H	> 4	11
J_H	6 + 3 pseudo	4 +1 pseudo
J_K	5	4+1 pseudo
J_λ	8	4
C_H	9 + 2 pseudo	8
C_K	1	1
C_λ	6 + 4 pseudo	4

Sequence of Immunoglobulin Gene Rearrangement

All cells in the body contain Ig genes in the germline configuration except B-lymphocytes. B cells express these genes in a functionally rearranged form and hence are the only cells that can synthesize the functional proteins. DNA rearrangement in B cells occurs in a highly regulated manner. It starts in the heavy chain locus by the joining of one D and one J segment with deletion of the intervening DNA (**Figure 6.3**).

This probably occurs even before lymphoid precursor cells are committed to the B cell lineage since DJ rearranged sequences are seen in some T cell tumours. The next step in the process is the joining of the DJ complex to one of the many V genes giving rise to a VDJ segment, with the deletion of intervening region including all D segments 5' of the rearranged D. An intron separates this VDJ complex from the 3′C region genes and this organisation is seen in the primary RNA transcript. Subsequently, this intron is spliced out, resulting in functional mRNA for the μ heavy chain since this is the most proximal C gene. Several consensus polyadenylation sites are located 3' of the

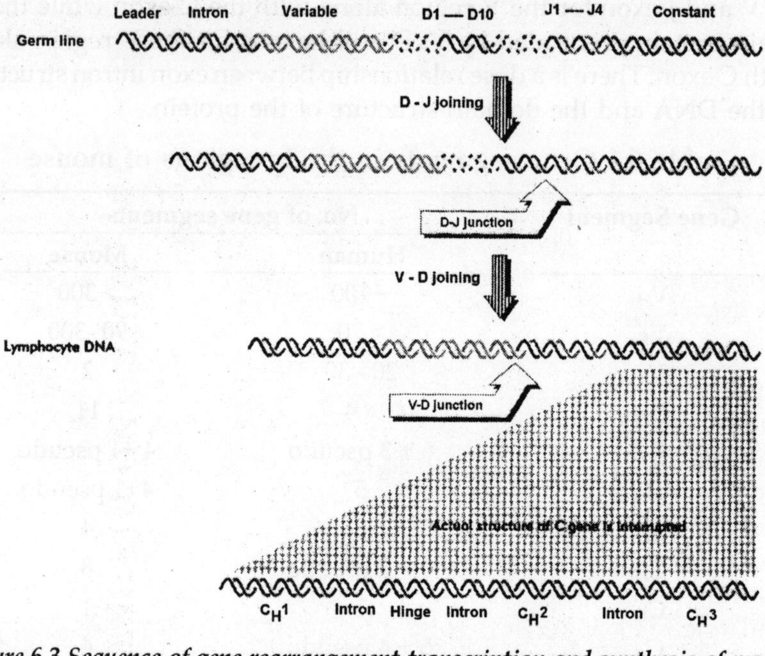

Figure 6.3 Sequence of gene rearrangement transcription and synthesis of mouse Igμ heavy chain

C RNA and several adenine nucleotides are added to one of these sites called poly-A tails. Other C chain genes also have their own 3' polyadenylation sites. Translation of the mRNA results in production of the "cytoplasmic" protein, the phenotypic marker of the pre-B cell.

DNA recombination is initiated in one of the light chain loci, and is essentially same as for heavy chain except that light chain genes do not have D segment (**Figure 6.4**). Hence the primary RNA transcript follows VJ rearrangement. Subsequent intron deletion and C gene joining to the VJ complex result in mRNA for k or λ protein. The light chain and previously synthesised heavy chain assemble together in the ER to form the membrane IgM molecule.

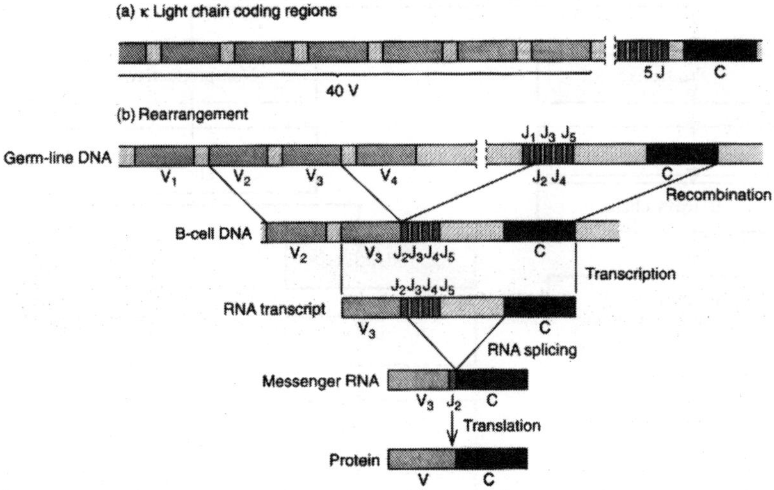

Figure 6.4 Sequence of gene rearrangement, transcription and synthesis of mouse Igk light chain

The μ chain protein in the developing B cell plays two important functions in the somatic recombination of Ig genes. The μ protein is produced by rearranged gene only on one chromosome, a phenomenon termed **allelic exclusion**. In other words, only one of the two loci of Ig on B cells are productively rearranged. In case both loci are non-productive, the cell undergoes death. The μ protein in pre-B cells also provides the signal for initiation of light chain rearrangement. The *k* light chains rearrange first in the pre-B cells and once this occurs, rearrangement in the λ chain is blocked. Rearrangement in the λ chain will occur only if the rearranged genes on both parental chromosomes fail to code for a functional protein.

This explains the observation that in any B cell only k or λ chains are produced; in k producing B cells, the λ chain genes are in germline configuration while in λ producing B cells k chain genes are either deleted or aberrantly recombined. Allelic exclusion works in light chain genes just as it does in the heavy chain genes and B cells die if neither of the light chain genes are productively rearranged. (**Figure 6.5**)

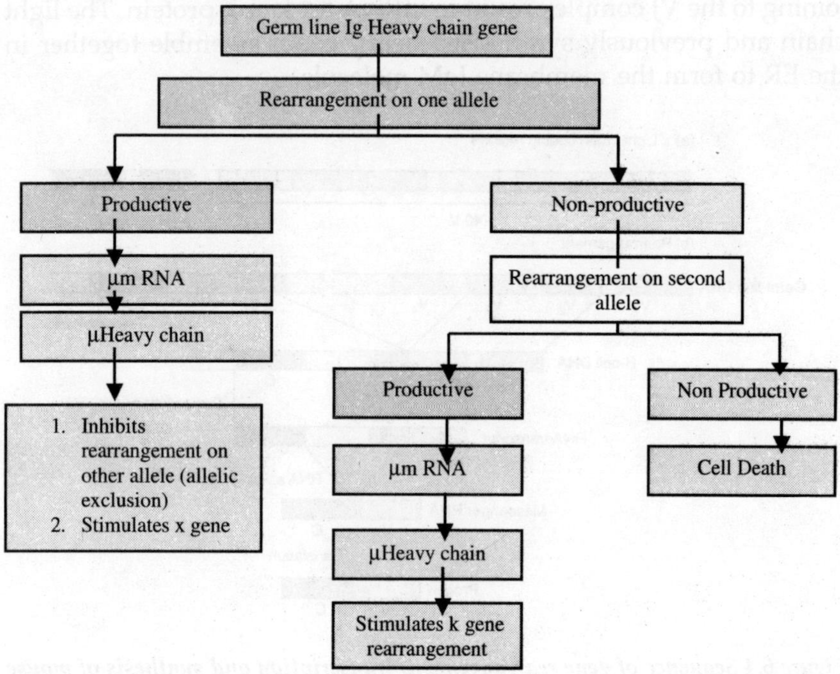

Figure 6.5 Order of rearrangement and expression of Ig heavy chain (μ) gene

Mechanisms of Immunoglobulin Gene Rearrangement

Ig gene rearrangement occurs through a special kind of recombination involving non-homologous gene segments. These events involve the deletion of DNA segments, between the genes that are to join together resulting in a continuous transcript for heavy and light chains. The recombinase enzymes perform this function. This enzyme is lymphocyte specific and recognizes specific sequences of DNA, which are highly conserved stretches of seven or nine nucleotides separated by non-conserved 12 or 23 nucleotide spacers (**Figure 6.6**).

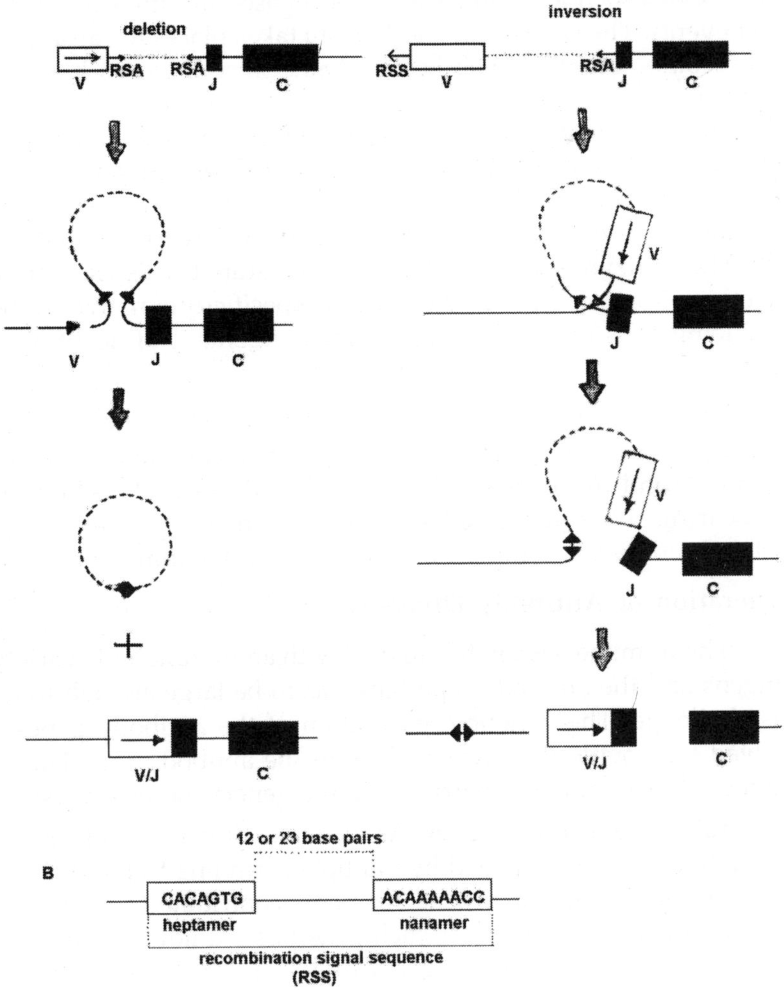

Figure 6.6 Mechanism of immunoglobulin k rearrangement: *The organization of variable (V), joining (J), and constant (C) DNA sequences within the genome is shown. Intervening segment of DNA forms a loop. Adjoining ends of the V and J exons are annealed and cause DJ and VDJ recombinations in the heavy chain*

These recognition sequences are located 3' of each V exon, 5' of each J segment and flanking both sides of each D segment. Thus the 3' end of V region and 5' end of J segment are brought together in the light chain and the intervening segment of DNA forms a loop which is removed, bringing adjoining ends of the V and J exons together to anneal and complete the process. Enzymes, which are not specific to lymphocytes such as exonucleases and ligases, are involved in these latter events. The same basic mechanism takes places to cause DJ and VDJ recombination's in the heavy chain. Since D genes have to be approximated to both V and J segments on either side, it has recognition sequences on both 5' and 3' ends. The same mechanisms are also operative in TCR chain rearrangement in T cells. Since recombinase enzyme is lymphocyte specific, it is not clear, why Ig rearrangement occurs only in B cells and TCR rearrangement in T cells. These enzymes are not functional in mature B cells and antibody secreting plasma cells. Thus the antibody specificity of native antibody producing cells cannot change by rearrangement of additional Ig genes.

Two genes, recombination-activating genes 1 and 2 (RAG-1, RAG-2), which as the name implies activate recombinase have been recently found in pre-B and immature T cells. Gene knockout mice not bearing these genes fail to produce both Ig & T cell receptor proteins and have a complete failure to produce mature T and B cells.

Generation of Antibody Diversity

The immune system has to deal with an innumerable variety of antigens and the antibody repertoire has to be large enough to meet this challenge. The genetic organization of the antibody genes has evolved to provide the diversity within the antibody population. A number of mechanisms contribute to the generation of diversity.

1. **Multiple germ line genes**: As we have already seen heavy and light chains are encoded by multiple germ line V, J and D genes (**Table 6.2**). All three gene segments contribute to the antigen-binding region and since each gene has a unique structure, the utilization of different germ line genes results in different antibody specificities. The degree of diversity that is generated at a particular Ig locus correlates with the number of gene segments.
2. **Combinatorial diversity**: the combination of different V, D and J segments adds a great potential for generation of antibody diversity, which is equal to the product of the number of V, D and J exons at each locus.
3. **Junctional diversity**: Even the same set of germ line V, D and J gene segments can generate different amino acid sequences at the junctional region by (a) combination of VJ or VDJ genes and (b) N region diversification by which nucleotides are added at

the junctional site by an enzyme terminal deoxytransferase.

4. Combination of heavy and light chain proteins.
5. Somatic mutations.

Table 6.2: Mechanisms Contributing to the Generation of Primary Antibody Diversity in the Mouse.

	H	κ	λ
Germ line Genes			
V gene segments	250-1000	250	2
J segments	4	4	3
D segments	12	0	0
Combinational Joining			
V X J (XD)	10,000 -40,000	1000	6
H-L Chain Associations			
HX	1-4 X 10^7		
HX	5-10 X 10^4		

All these mechanisms operate in the primary or naïve B cell repertoire. Thus, during the primary response to a complex antigen, different B cell clones expressing different VJL and VDJH combinations will be stimulated. But during a secondary response to the same antigen, only a few of these clones will expand and clones with highest affinity for the antigen will be activated.

Antibodies produced during the secondary response have a much greater affinity for the antigen then those produced during the primary response. Increased affinity is due to small changes in amino acid sequence of variable regions of the heavy and light antibody chains. The change in sequence results from mutations in the genes that encode this polypeptide. It is estimated that rearranged DNA elements encoding antibody V regions have a mutation rate 10^5 times greater than the other genetic loci in the same cell. The mechanism responsible for this increased level of V-region mutation is not understood, but it is thought to involve some form of error prone DNA repair. Regardless, those B cells that carry mutations producing Ig molecules with greater antigen affinity are preferentially selected following antigen re-exposure. Selected cells proliferate to form clones that undergo additional rounds of somatic mutation and selection, whereas non-selected cells that express low affinity Ig molecules undergo apoptosis. In this way, the antibody response to recurrent or chronic infections improves over time.

Expression of different Classes and Types of Immunoglobulins

The earliest Ig to be produced is the cell surface expressed IgM on pre-B cells and consists of one μ heavy chain coupled to either k or λ light chain. As the cell matures, other than mutations, further changes do not take place in the VDJ segments of the Ig molecule. The constant regions of heavy chains undergo changes in gene expression to give secreted forms of IgM as well as of other isotypes.

The mature B-lymphocytes express both IgM and IgD. Both the μ and δ chains on a cell have the same V region and hence same antigen specificity. The expression of the same VH with both Cμ and Cδ to form 2 heavy chains occurs by alternate splicing of a primary RNA transcript consisting of the rearranged VDJ complex as well as sequences for Cμ and Cδ.

Membrane Bound and Secretary Ig

Antibody molecules exist as membrane bound molecules on the surface of cells and as soluble secreted molecules in the plasma or other body fluids. Which of the two forms an immunoglobulin will assume is determined by the amino acid sequence at the 3′ end of the heavy chain constant region. The molecular basis for this is evident from the organisation of the constant region exons.

The heavy chains of membrane Igs have a slightly higher molecular weight than their secreted counter parts as they contain additional aminoacid residues which form the hydrophobic transmembrane portion and exons encoding the cytoplasmic tail of the molecule. The exon or exons encoding the transmembrane regions and the cytoplasmic tail are located downstream of C 4 exons (**Figure 6.7**). These exons encode 26 hydrophilic cytoplasmic tail of variable length being 3 amino acid long in μ chain but 28 amino acid long in γ chains. Regulation of the production of membrane Ig versus secreted immunoglobulin is probably controlled at the RNA level and involves RNA processing. Indeed two different species of mRNA, one of 2.7 kb (membrane) and the other of 2.4 kb (secreted) can be detected in B cells producing IgM.

In all cases, transmembrane regions of membrane Ig heavy chains show a high degree of homology. They all adopt the conformation of helix as they span the lipid bilayer.

Class Switching

The type of CH region determines the isotype of the immunoglobulin. All lymphocytes start productive life as immature cells engaged in synthesis of IgM. As the response matures, the cell switches to produce one of the other isotypes a phenomenon called class switching.

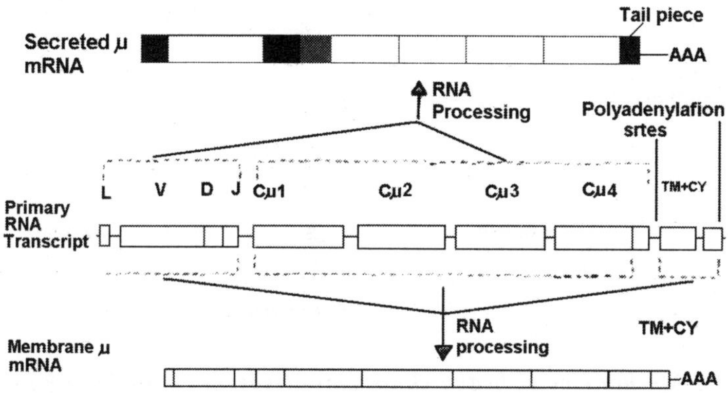

Figure 6.7 Expression of membrane and secreted μ chains by B lymphocytes.

Switching involves only the CH genes as the B cell continues to express the same light chain throughout the lineage of the cell. Class switching therefore allows the type of effector response (mediated by the CH region) to change, while maintaining a constant facility to recognize antigen (mediated by the V - region).

Cells expressing later CH genes have deletion of C and other genes preceding the expressed CH gene. Class switching, which is accomplished by a recombination, brings a new CH gene into juxtra position with the expressed V-D-J unit (**Figure 6.8**). The sequence of switched V-D-J-C unit shows that the sites of switching lie upstream of the CH genes themselves; the switching sites are called switch (S) regions.

The S region, varies in length from 1 to 10 kb, and contain group of short homologous repeats each of which is up to 52 bp long. The recombination reaction is not site-specific, but usually occurs near the conserved repeat elements.

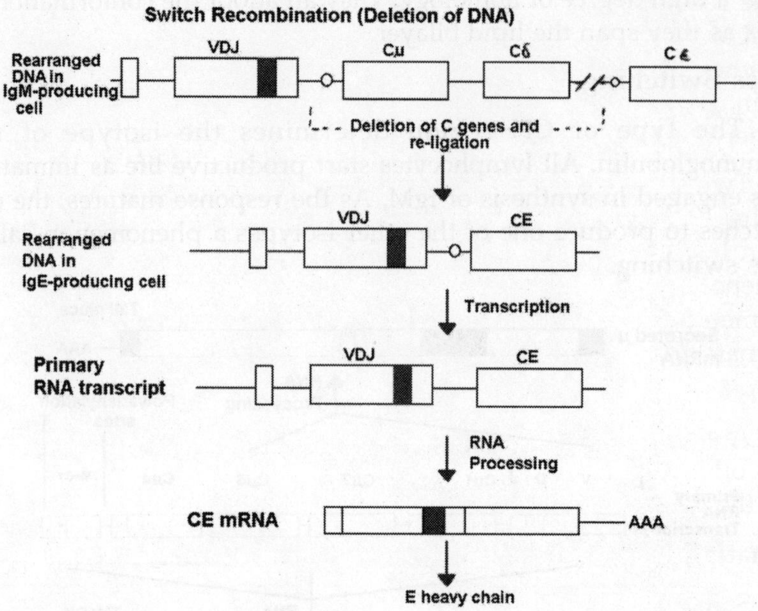

Figure 6.8 Mechanism of heavy chain class (isotype) switching by switch recombination

The molecular basis of the reaction is not well understood. The excised material between the switch sites is released as a circular DNA molecule. Because the S region lies within the introns that precede the CH coding regions, switching does not alter the transitional reading frame.

The linear deletion model imposes a restriction on the heavy chain gene locus. Once a class switch has been made, it becomes impossible to express any CH gene that lies between Cμ and the new CH gene. However, it is possible to undertake another switch to any CH gene downstream of the expressed gene.

The co-existence of productive and non-productive rearrangements suggests that existence of a feedback loop to control the recombination process. Let us suppose that each cell starts with two loci in the unarranged germ line configuration IgD. Either of these loci may be rearranged to generate a productive gene (Ig+) or non-productive gene (Ig-).

If the rearrangement is productive, the synthesis of an active chain provides a trigger to prevent rearrangement of the other allele.

If the rearrangement is non-productive it creates a cell with the configuration Ig0 / Ig-. There is no impediment to rearrangement of the other allele and the expressing cell has the configuration Ig+ / Ig. Again, the presence of an active chain suppresses the possibility of further rearrangements.

Two successive non-productive rearrangements produce the cell Ig-/Ig-. In some cases an Ig-/Ig- cell can try yet again. The chain ends when hydrophobic sequence is replaced by a shorter hydrophilic sequence and the substitution allows the heavy chain to pass through the membrane; the change of C terminus is accomplished by an alternate splicing event, which is controlled by the 5' end of the nuclear RNA.

A similar transition from membrane to secreted forms is found with other constant regions. Conservation of exon structure suggests that the mechanism is the same.

Somatic Mutation Generate Additional Diversity

Somatic mutations can occur in mature lymphocytes following complete rearrangements and add to the degree of diversity. Mutation occurs at random in V region or beyond it and there are certain hot spots where mutation occurs frequently. Molecular basis are not known for such mutations. They may be induced by antigenic stimulation and occur during clonal proliferation apparently at a rate of more than 10^{-5} per bp per cell generation.

There are many cases in which a single family of V genes is used consistently to respond to a particular antigen.

Somatic mutation creates B cells, which have higher affinity for antigens. During subsequent cycles of cell activation, antigen binds to and selects cell expressing the receptor with higher affinity. Thus, as the immune response matures the antibody structure is fine-tuned to fit better and better into the antigen. This is termed **affinity maturation**.

CHAPTER 7

T cell Receptor Complex

T cells have specific receptors for antigen fragments (e.g., peptides with a linear sequence of 8 to 25 amino acids) on their plasma membrane surface. The receptor site is composed of two parts, an alpha polypeptide chain and a beta polypeptide chain (**Figure 7.1**). Each chain is stabilized by disulfide bridges. The receptor is anchored in the plasma membrane by parts of the α and β chains that extend into the cytoplasm. The recognition sites of the T cell receptor extend out from the membrane and have a terminal variable section complementary to antigen fragments. T cells respond to antigen fragments exposed on the surfaces of antigen-presenting cells.

The central point of an effective immune response is the trimolecular complex made of the major histocompatibility complex (MHC) molecule, the antigenic peptide and the T cell receptor (TCR) (**Figure 7.1**).

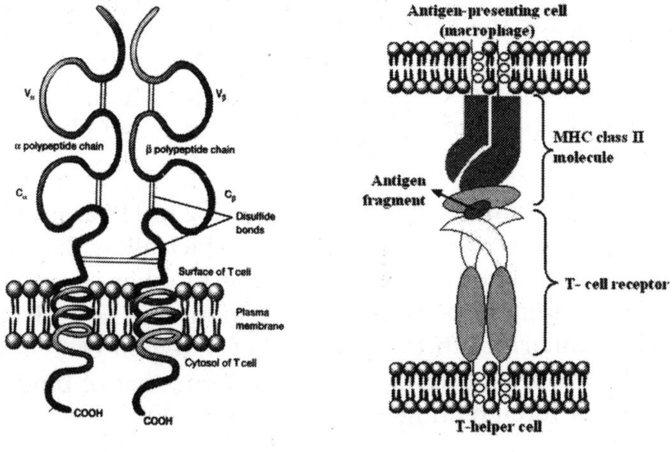

Figure 7.1 ***T-cell receptor protein and its activation.*** *(a) A schematic illustration of the proposed overall structure of the antigen receptor site on the T-cell plasma membrane. (b) An antigen-presenting cell begins the activation process by displaying the antigen fragment on its surface as part of a complex with the histocompatibility molecules. A T-helper cell is triggered when the variable region of its T cell receptor (designated $V\alpha$ and $V\beta$) reacts with the antigen fragment on the presenting cell surface.*

The recognition of the MHC -peptide complex by the TCR results in T cell activation with consequent secretion of cytokine and proliferation and performance of regulatory or cytolytic effector functions by the appropriate T cell type. The TCR is composed of a complex of several integral plasma membrane proteins, consisting of those that are directly involved in specific binding to the peptide - MHC complex and others that are invariant among the T cell subpopulations and transmit the signals from cell surface to the interior. In addition to this complex of proteins, T cells also express a number of other cell surface proteins, which are called accessory molecules. These latter molecules are essential for the full functional activity of the T cells.

The αβ T cell receptor

The receptor on the majority of T cells for the peptide -MHC complex is the αβ heterodimer, which consists of two covalently linked polypeptide chains linked to one another by disulphide bonds. A second alternate type of TCR, composed of γ δ chains has a limited distribution on small subpopulations of T cells and will be discussed later. The α and β chains are 40 to 60 kilo Dalton (kD) respectively and are uncharged or basic glycoproteins. Both these chains have variable (V) and constant (C) regions with a J or joining segment gene encoding for the carboxyl terminal end of the V region. In addition, the β chain also has a diversity (D) segment gene. The presence of V, D, J and C regions and their structural arrangement are very similar to that of the immunoglobulin V and H chain genes (**Figure 7.2, 7.3**).

The V region of the α and β chains are 102 to 119 amino acids in length. They contain two cysteine residues, which allow the formation of an intra chain disulfide bond. A large number of amino acid residues are conserved between the V region of TCR and Ig super family. The similarities in the structure of V region of TCR and Ig may reflect the common function they perform, that of binding to a large variety of foreign antigens.

The C region of the α and β chains are 138 to 179 amino acids in length and consist of four functional domains that are encoded by separate exons. The most amino terminal domain contains two cysteine residues, which form the intra chain disulfide bonded loop. The second domain is the hinge region and contains a cysteine residue involved in the inter chain disulfide linkage. The third part of the C chain is

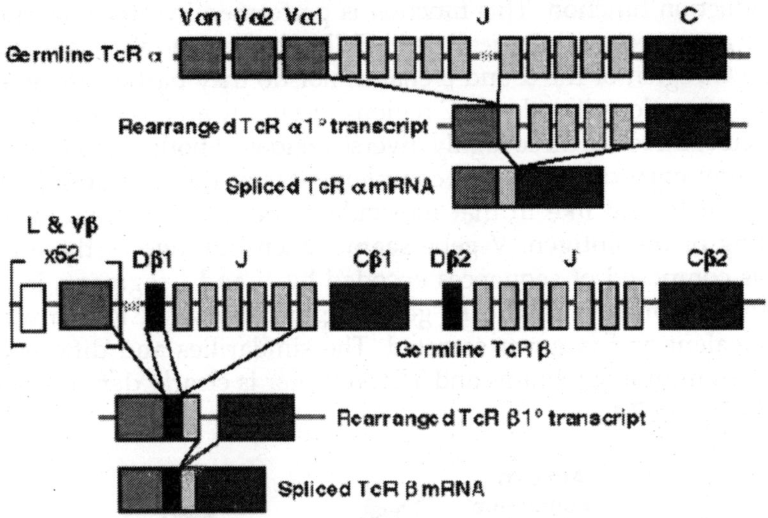

Figure 7.2 Organization of human T cell receptor (TCR) genes in germ line

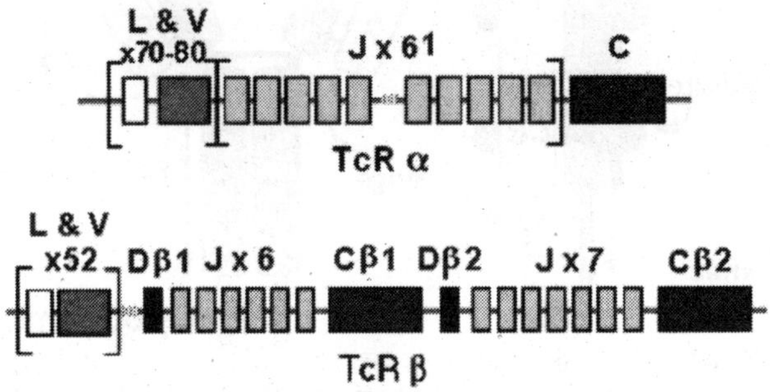

Figure 7.3 Organization of mouse T cell receptor (TCR) genes in germ line. *Segmentation of TcR genes into V, D, J and C (Variable, Diversity, joining and constant) resembles Ig genes.*

the transmembrane domain; it consists of 20 to 24 amino acid residues, which are predominantly hydrophobic. It also contains a lysine residue in the β chain, the positively charged side chains, which may be important for interaction with the negatively charged residues in the transmembrane portion of the CD3 molecule. The fourth domain or the carboxy terminal portion of the α and β chains is formed by 5 to 12 amino acid long cytoplasmic tail. We have seen that the main function of the TCR is T cell activation that follows its engagement by the appropriate peptide -MHC complex. However, the cytoplasmic tail of the TCR heterodimer is too small to serve an intrinsic signal transduction function. This function is performed by other peptides, which are associated physically with TCR. There are no alternate forms of the C region of the α and β chains; nor do they participate in any effector functions like the Ig C regions. Again by analogy with the Ig molecule, there are three highly diverse regions in both α and β chains that may correspond to the complementarily determining region (CDR) of Ig and like in that molecule form contact points for the binding of the antigen. V gene segments encode two of these and one is composed of sequences encoded by V and J segments in the chain. In contrast to the Ig gene super family, TCR is always monovalent and is never secreted. The similarities and differences between immunoglobulins and TCR receptor is clearly demonstrated in the **Figure 7.4.**

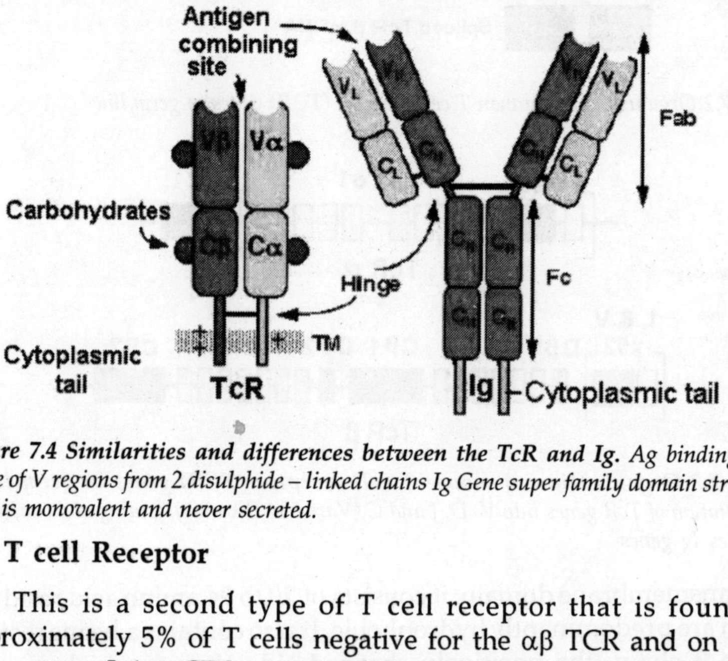

Figure 7.4 Similarities and differences between the TcR and Ig. *Ag binding sites made of V regions from 2 disulphide – linked chains Ig Gene super family domain structure TcR is monovalent and never secreted.*

γ δ T cell Receptor

This is a second type of T cell receptor that is found on approximately 5% of T cells negative for the αβ TCR and on some thymocytes. It is a CD3 associated disulfide linked heterodimer, with the γ and δ chains having close structural similarity to the αβ chains. Both chains have the extracellular Ig like V and C regions, short connecting or hinge region, hydrophobic transmembrane segments and short intracytoplasmic tails. The transmembrane segments of both γ and δ contain positively charged lysine residues and the δ chain contains an additional positively charged arginine residue. These help in the close covalent association of these proteins with the CD3 polypeptides. In their primary amino acid sequence the γ chain is similar to that of the TCR β chain while the δ chain sequence is like that of the α chain. The γ and δ chains are glycoproteins of 36 to 55 kD and 40 to 60 kD respectively.

The γδ cells are predominantly found in epithelial locations. Although their arrangements are similar to that of the αβ receptor, the diversity that is generated is much less than in the latter. Unlike the αβ receptors which exhibit a tremendous diversity and are functionally evolved to recognise a large number of extra cellular antigens, the γδ receptors by their restricted diversity seem to have evolved for recognition of less polymorphic legends. Thus, it has been postulated that they play a role in the primary defence of the body by recognising and dealing with a limited number of frequently encountered antigens at epithelial surfaces.

T cell receptor genes

Functional TCR α and β chain genes are present only in cells of T cell lineage and are formed by somatic rearrangements of germline gene segments by a process that is very similar to Ig gene rearrangement. The gene for the human α chain is located on chromosome 14 and for the β chain on chromosome 7. Of the number V, J, D and C gene segments for the α and β chains, the number of V region segments are fewer than in the Ig heavy and light chains while the number of J region genes are far more (**Table 7.1**). The implications

Table 7.1 The numbers of human T-cell receptor gene segments and the sources of T-cell receptor diversity compared with those of immunoglobulins.

Element	Immunoglobulin		□: □ T-cell receptors	
	H	κ + λ	□	□
Variable segments (V)	40	70	52	~ 70
Diversity segments (D)	25	0	2	0
D segments read in three frames	Rarely	-	Often	-
Joining segments (J)	6	5(κ) 4 (λ)	13	61
Joints with N- and P- nucleotides	2	50% of joints	2	1
Number of V gene pairs	1.9×10^6		5.8×10^6	
Junctional diversity	$\sim 3 \times 10^7$		$\sim 2 \times 10^{11}$	
Total diversity	$\sim 5 \times 10^{13}$		$\sim 10^{18}$	

of this are not understood. However, the overall diversity of the TCR has been calculated to be many folds higher than for the immunoglobulin.

During T cell maturation in the thymus, the gene segments are rearranged in a defined order resulting in functional α and β genes with ordered V, D, J and C segments lying in close proximity to one another. The α chain locus rearranges before the β locus and begins with the joining of one-D segment with one-J segment. This newly formed DJ segment is then joined to a V segment to give the VDJ gene. In the β chain however, direct VJ joining can occur unlike in Ig genes. The enzymes involved in mediating rearrangements (recombinse) as well as the recognition sequences are the same as has been previously described for Ig genes. Recombination takes place by looping out and deletion as illustrated in the chapter on immunoglobulin. The primary transcripts of the TCR β genes contain non-coding sequences between VDJ and C genes, which are spliced out to form the mature RNA. However, there are only two C segments for β genes and the VDJ segment gets associated at random with either of these. There are no isotypic variants of the TCR and no evidence that any form of switch from one Cβ gene to another takes place. In this respect the TCR rearrangement is different from that of Ig.

Functional rearrangement and possibly transcription of β chain genes appear to provide the signal for initiation of α chain gene rearrangement. The process is similar to that of the β chain gene except that the α chain has no D segments and V segment directly joins to J segment to initiate transcription.

Although the mechanisms involved in TCR and Ig rearrangements are the same, Ig genes are functionally rearranged only in B cells and TCR genes are rearranged only in T cells. The basis for this lineage specificity is not understood. Promoter regions have been identified 5' of the V segments of both the Cβ and Cα as well enhancer regions 3' of the Cβ and Cα genes. Whereas the 5' promoters are not T cell specific and are only minimally active, they are responsible for high level T cell specific activity when the 3' enhancer is brought in approximation. Not much is known about the nuclear binding proteins or consensus binding sequences, which are involved in the regulation of TCR genes.

Allelic exclusion takes place in TCR genes just as it does in Ig and the successful rearrangement of one locus shuts down rearrangement at the other locus; failure or defective rearrangement at both loci results in the death of the cell. The mechanisms, which generate diversity, are also the same as in Ig.

The γ chain locus is located on human chromosome 7 while the δ chain is unusual as it is located entirely within the α chain locus between the Vα gene segments and JC α clusters. The rearrangements of the two chains are similar to what has been discussed for the αβ chain.

The CD3 ξ and η protein

As we have already seen, TCR is the crucial molecule on T cell for antigen recognition. However, it is unable through its αβ chains to transmit the intracellular signals that are needed for T cells to get activated and perform effector functions. This important function is performed by the proteins that are non-covalently linked with the αβ heterodimer and together this forms the functional TCR complex. The four additional chains are the 2 ε chains one δ and one γ chain of the CD3 molecule and homodimer of ξ.

The CD3 molecule consists of a 25 to 28 kD glycosylated γ chain, a 20 kD glycosylated δ chain and a 20 kD non glycosylated ε chain. In addition, the majority of the TCR complexes contain a homodimer of 16 kD non glycosylated ξ chain.

The genes for the γ, δ and ε chains are homologous and located on chromosome 11. Each of these proteins contains an N- terminal extracellular region, a short connecting peptide, a transmembrane region and an intra cytoplasmic tail. These proteins are members of the Ig super gene family since their extracellular domain contain one Ig like domain. These proteins are invariant in nature (i.e. have no variability or polymorphism of structure) and play no role in the actual recognition of antigen. The transmembrane portion has a negatively charged aspartic acid residue which is important for its physical and functional association with the α and β chains of the TCR. As we have already seen, these latter molecules bear negatively charged residues in their transmembrane segments. The cytoplasmic tail of the γ, δ and ξ proteins vary from 44 to 81 amino acids in length. This is sufficiently long to transduce signals to the cell interior. The cytoplasmic tail of each of these proteins has a conserved sequence motif, which is also present in the cytoplasmic tail of Ig, Ig proteins associated with membrane IgM and IgD as well as the end chains of the TCR complex. This motif is called the immuno receptor tyrosine based activation motif (ITAM) and all the molecules, which bear this motif, are involved in signal transduction.

The ξ and η chains are products of the same gene on chromosome 1 and are produced by alternate splicing mechanism. Their extracellular domains consist of only nine amino acids and both these domains and the transmembrane domain of the two chains are identical in their amino acid sequence. The ξ and η differ only in their long (113 to 155 amino acids) cytoplasmic tails.

It should be appreciated that the CD3 and ξ / η proteins are expressed in very early immature T lymphocytes prior to TCR α and β chain expression. The protein products of the CD3 genes are post translationally modified and form structures in the absence of the TCR chains. The association of αβ heterodimer with the CD3 complex occurs in the endoplasmic reticulum (ER). From here the complex is transported to the Golgi apparatus, where further modifications at the glycosylation sites takes place. Incomplete molecules are not transported to the cell surface. The TCR α, β and CD3 γ, δ and ξ chains are synthesized in excess; since the association with homodimer is essential for the surface transport it seems that the synthesis is the rate limiting step in assembly and surface expression of TCR complex.

Thymic selection

Progenitors of human T cells migrate from the bone marrow to the thymus in the 8th to 9th week of development. In the thymus T cells gain their identity or maturate. During this process rearrangements of germ line TCR (T cell receptor) genes and expression of various membrane molecules takes place. In the thymus T cells are selected to recognize self-MHC molecules and T cells that do not recognize self-MHC molecules and that recognize self-antigens are eliminated. T cells gaining maturation in the thymus are known as thymocytes. Initially thymocytes do not have any of the characteristic features of T cells i.e. they do not express CD4 or CD8 and are termed double negative. The first membrane molecule to appear is c kit, a receptor for stem cell growth factor, followed by CD44, which is a cell adhesion molecule. Finally, CD25, which is the α chain of the interleukin - 2 (IL2) receptor, is expressed in the developing thymocytes. These thymocytes are able to proliferate but have not rearranged their TCR genes. Thymocytes then stop expressing c- kit, reduce the expression of CD44, and start to express the RAG_1 and RAG_2 and begin to rearrange the TCR genes. A small fraction of thymocytes productivity rearranges γ and δ gene segments and they develop into double negative CD3 + γδ T cells. These cells represent more than 80% of CD3+ cells in early gestation and then decline in number to reach less than 50% in adults.

Majority of the double negative cells stop proliferating and start to rearrange and express the TCR β chain. Initially, the β chain associates with a glycoprotein called pre-Tα chain to form the pre- T cells receptor (pre-TCR), which associates with the CD3 complex. The interaction of pre-TCR with an unknown ligand sends a signal through the CD3 complex that activates signal transduction proteins, like lck, Zap -70 and syk. These signals suppress further rearrangements of β chain genes resulting on allelic exclusion of the β chain. Also there is induction of TCR α chain gene rearrangements and thymocyte start to express both CD4 and CD8 and hence are called double positive (DP) cells. DP cells stop proliferating as α chain rearrangements is initiated. The DP cells with αβ TCR and CD3 complex develop into either single positive CD4+ or single positive CD8+. Positive selection takes place in the cortex of the thymus and involves the interaction of thymocytes and cortical epithelial cells. Only cells whose αβ TCR recognizes a self-MHC molecule are selected for survival and cells that do not receive the survival signal die by apoptosis. This selective process generates self- MHC restricted T cells. In the thymic medulla thymocytes bearing TCR with high affinity for self- antigens plus self-MHC or for self-MHC molecules alone, are eliminated by apoptosis. This negative selection ensures that the mature immune system does not get activated by the self-antigens, a mechanism termed central tolerance.

Interaction of the TCR with MHC molecule

Recent crystallographic studies have provided a striking view of how the external regions of the TCR bind to peptide + MHC complexes. The CDR3 regions of the TCR α and β chains, the regions of most variability in the TCR, interact with the centre of the peptide bound in the groove of the MHC molecule. This central area of the peptide gets out of the MHC groove and is the only part of the peptide not buried in the MHC molecule. This suggests that limited contacts with amino acids in the centre of the peptide are critical for recognition by the TCR. In contrast, the majority of the TCR-MHC interactions are with conserved MHC amino acids for that allele. In particular, the CDR2 regions of the TCR α and β make a number of contacts with the helical walls of the MHC molecule, "anchoring" the interaction of MHC and TCR. Thus, the TCR appears to make different types of contacts with the peptide on one hand and the MHC molecule on the other. The TCR binds diagonally across the groove of the MHC molecule; this orientation is thought to maximize the interactions between TCR, peptide, and MHC class I molecules. The interaction of TCRs with MHC class II molecules is similar.

Lymphocytes are activated by cell -surface signals

Lymphocytes communicate with other cells through an array of cell-surface proteins. T-cell activation requires an interaction between the TCR of the T cell and an MHC-peptide complex on the surface of another cell. This interaction provides specificity, and ensures that only T cells that bind the antigen are activated. T-cell activation also requires a second signal, called the co-stimulatory signal, which is delivered through a second type of receptor on the surface of a T cell. This receptor is distinct and spatially separate from the TCR. Unlike the TCR, the receptor that delivers the co-stimulatory signal is not specific for a particular antigen and does not require an MHC molecule to bind. The best studied of these interactions is between helper T cells and professional antigen presenting cells (e.g., dendritic cells and macrophages).

Activation of helper T cells by professional APCs

T_H cells recognize antigen fragments on the surface of dendritic cells and macrophages that are lodged in the binding cleft of MHC class II molecules. A co-stimulatory signal is delivered to a T_H cell as the result of an interaction between a protein known as CD28 on the surface of the T_H cell and a member of the B7 family of proteins on the surface of the APC. T_H cells that do not receive the second signal from the APC, either become non-responsive (anergized) or are stimulated to undergo apoptosis (deleted). Because professional APCs are the only cells capable of delivering the co-stimulatory signal, they are the only cells that can initiate a T cell response. As a result, other cells of the body that may bear peptides on the MHC II are capable of recognition by the TCRs but fail to activate the T cell, as they lack the co stimulation molecule. This requirement by a T_H cell for two activation signals is a clever protective mechanism that protects normal cells from autoimmune attack.

Prior to its interaction with an APC, a T_H cell is a resting cell that has withdrawn from the cell cycle (a G_0 cell). Once it receives the dual activation signals, it is stimulated to re-enter the G_1 phase of the cell cycle and eventually to progress through S phase into mitosis. Thus, interaction of T cells with a specific antigen leads to the proliferation (clonal expansion) of those cells capable of responding to that antigen. In addition to triggering cell division, activation of a T_H cell causes it to synthesize and secrete cytokines (most notably IL-2). Cytokines produced by activated T_H cells act on other cells of the immune system (including B cells and macrophages) and also on the T_H cells that secrete the molecules. The source and function of some important cytokines are shown in **Table 7.2**.

Table 7.2 Selected Cytokines

Cytokine	Source	Major functions
IL-1	Diverse	Induces inflammation, stimulates T_H-cell proliferation
IL-2	T_H cells	Stimulates T-cell and B-cell proliferation
IL-4	T cells	Induces IgM to IgG class switching in B cells, suppresses inflammatory cytokine action
IL-5	T_H cells	Stimulates B-cell differentiation
IL-10	T cells macrophages	Inhibits macrophage function, suppresses inflammatory cytokine action
IFN-γ	T_H cells, CTLs	Induces MHC expression in APCs, activates NK cells
TNF-α	Diverse	Induces inflammation, activates NO production in macrophages
GM-CSF	T_H cells, CTLs	Stimulates growth and proliferation of granulocytes and macrophages

Ligand receptor interaction initiates a series of intracellular events, of both activating and inhibitory in nature; a balance of positive and negative influences determines the ultimate response by the cell. For example, interaction between CD28 and a B7 protein delivers a positive signal to the T cell leading to its activation. Once the T cell has been activated, it produces another cell-surface protein called CTLA4 that is similar to CD28 in structure and also interacts with B7 proteins on APC. CTLA4-B7 interaction leads to inhibition of the T cell's response, rather than activation. The role of the inhibitory interaction is seen in mice that have been genetically engineered, to knockout the gene encoding CTLA4. These mice die as a result of massive over proliferation of T cells. Soluble forms of CTLA4, capable of interacting with B7 receptors on the surface of APCs and leading to inactivation of T cells are being used in clinical situations.

Activation of B cells by T_H cells

T_H cells bind to B cells whose receptors recognize the same antigen. Antigen binds to the immunoglobulin receptor (BCR) on the B-cell surface and is internalised by the B cell, where it is processed enzymatically. The peptide fragments are displayed in combination with MHC class II molecules as in any other antigens presenting cell. B cell activation follows transmission of several signals from the T_H cell to the B cell. Some of these signals are transmitted directly from one cell surface to the other through an interaction between complementary proteins, such as CD40 and CD40 ligend (CD40L). Binding between CD40 and CD40L generates signals that help to move the B cell from the G_0 resting state back into the cell cycle. Other signals are transmitted by cytokines released by the T cell into the space separating it from the nearby B cell. This process is not unlike the way neurotransmitters act across a neural synapse. Cytokines released by the T cells into the "immunological synapse" include IL-4, IL-5, IL-6, and IL-10. Interleukin 4 is thought to stimulate B cell to switch from producing the IgM class in producing the IgG or IgE class. Other cytokines induce the proliferation, differentiation, and secretary activities of B cells.

Signal transduction pathways used in lymphocyte activation

When a T cell is activated by contact with an MHC-peptide complex, a signal is transmitted from plasma membrane to cytoplasm by tyrosine kinases. Lymphocyte antigen receptors lack an inherent tyrosine kinase activity. Instead, ligand binding to antigen receptors leads to the recruitment of cytoplasmic tyrosine kinase molecules to the inner surface of the plasma membrane. Several different tyrosine kinases have been implicated in signal transduction during lymphocyte activation, including members of the src family (e.g., Lck and Fyn). Src was the first tyrosine kinase to be identified and is the product of the first cancer-causing oncogene to be discovered.

Activation of various tyrosine kinases lead to the activation of at least three distinct signal transduction pathways:

1. Phospholipase C, which leads to the formation of inositol-phosphate-3 (IP3) diacylglycerol (DAG). IP3 and causes a marked elevation in the levels of cytosolic Ca2+, whereas DAG stimulates protein kinase C activity.
2. Ras, which leads to the activation of the MAP kinase cascade.
3. Phosphatidylinositol 3-hydroxy kinase or P13K, which catalyses the formation of membrane bound lipid messengers having diverse functions in cells.

Transmission of signals along various pathways, leads to the transcriptional activation of dozens of genes that are not expressed in the resting T cell.

As noted above, one of the most important responses by an activated lymphocyte is the production and secretion of cytokines, some of which may act back on the cell that releases them. Like other extracellular signals, cytokines bind to receptors on the outer surface of target cells, generating cytoplasmic signals that act on various intracellular targets. Cytokines utilize a novel signal transduction pathway referred to as the JAK-STAT pathway. The "JAK" portion of the name is an acronym for Janus kinases, a family of cytoplasmic tyrosine kinases whose members become activated following the binding of a cytokine to a cell-surface receptor (Janus is a two-faced Roman god who protected entrances and doorways) "STAT" is an acronym for "signal transducers and activators of transcription", a family of transcription factors that become activated when one of their tyrosine residues is phosphorylated by a JAK. Once phosphorylated, STAT molecules interact to form dimmers that translocated from the cytoplasm to the nucleus where they bind to specific DNA sequences, such as an interferon-stimulated response element (ISRE). ISREs are found in the regulatory regions of a dozen or so genes that are activated when a cell is exposed to the cytokine interferon a (IFN-α).

The specific response of a cell depends on the particular cytokine receptor engaged and the particular JAKs and STATs that are present in that cell. For example, it was noted earlier that IL-4 acts to induce Ig class switching in B cells. This response follows the IL-4 induced phosphorylation of the transcription factor STAT6, which is present in the cytoplasm of activated B cells. Resistance to viral infection induced by interferons is mediated through the phosphorylation of STAT1. Phosphorylation of other STATs may lead to the progression of the target cell through the cell cycle.

CHAPTER 8

Complement System

The complement system is a complex group of interacting blood-derived proteins and glycoproteins found in all vertebrates. It forms a major effector mechanism in the humoral immune response and is involved in lysis of infectious agents. Seventeen component members of the complement system are involved in the activation and twelve in the regulation of the complement cascade. The functions of the complement system include initiation of inflammation, opsonization of foreign material for phagocytosis and direct cytotoxicity of various cells and micro-organisms. The various protein components that are involved in activation phase of complement system are shown in **Table 8.1**.

Table 8.1 Components of the Human Complement System

Component symbol	Concentration in plasma (μg/ml)	Molecular weight (Dalton)	Number of chains in native form	Associated pathway*	Enzymatic activity
Clq	180	462,000	18 (6ea A, B, C)	CP	No
Clr	100	92,000	1	CP	Yes
Cls	110	86,000	1	CP	Yes
C2	25	117,000	1	CP, MBL	Yes
C3	640	206,000	3	CP, MBL	Cofactor
C4	1200	185,000	2	CP, MBL,AP	Cofactor
C5	80	180,000	2	Terminal	Cofactor
C6	75	120,000	1	Terminal	Yes?
C7	55	110,000	1	Terminal	No
C8	80	163,000	3	Terminal	No
C9	50	71,000	1	Terminal	No
MBL (mannose-binding lectin)	0.1-5	540,000	18	MBL	No
$MASP_1$	ND	94,000	1	MBL	Yes

Component symbol	Concentration in plasma (□g/ml)	Molecular weight (Dalton)	Number of chains in native form	Associated pathway*	Enzymatic activity
MASP$_2$	ND	90,000	1	MBL	Yes
C3b or C3.H_2O	Trace	170,000	2	AP	Cofactor
D (factor D)	2	25,000	1	AP	Yes
B (factor B)	200	93,000	1	AP	Yes
P (properdin)	25	220,000		AP	No
Cl-inhibitor	25	110,000	1	CP control	No
Factor I (I. C3b-Ina, KAF)	35	88,000	1	Control, all pathways	Yes
H (factor H, □1H)	500	150,000	1	AP control	Cofactor
C4-binding protein	250	550,000	7	CP control	Cofactor
CFI (CPB-N)	35	310,000	1	Control	Yes
SP40,40	50	80,000	1	Terminal control	No
S (vitronectin)	500	83,000	1	Terminal control	No
C4a	1.6	12,000		CP split product	No
C4d	8.9	30,000		CP split product	No
C3a	0.6	11,000		TP split product	No
iC3b	8.5	170,000		TP split product	No
C5a	0.01	11,000		TP split product	No
Bb	0.4	60,000		AP split product	No
SC5b-9 TCC	0.3	>1,000,000		Terminal C complex	No

* CP, classical pathway; MBL, mannose-binding lectin pathway; AP, alternate pathway; TP, terminal pathway; MASP.

Three distinct pathways, all of which cause cell lysis, the classical, alternate and lectin pathways activate the complement cascade. Each of the individual components of the classical pathway has a prefix C while the components of alternate pathway are factor B, factor D and properdin.

The complement system has a number of important properties that enable it to function efficiently in host defence without causing injury to normal tissues. The soluble serum complement components are proteolytic enzymes, which are present in blood in an inactive form. They become activated only when they are cleaved by other previously activated enzymes in the cascade. The process of sequential activation, or enzymatic cascade, gives the system capacity for amplification since each individual enzyme molecule activated at one step can generate multiple activated enzyme molecules at the next step. An additional property of the complement system is tight counter-regulation which limits the activation by physiological stimulation and controls the response. The central component of the complement system is the protein, C3, which is cleaved by C3 convertase to C3a and C3b (by convention, lower case letters denote proteolytic products). The different pathways of complement activation differ essentially in their requirements for generation of distinct proteins with C3 convertase activity. Once C3 is cleaved, subsequent pathway is similar and results in the generation of the membrane attack complex (MAC).

The Classical Pathway

Initiation of the activation process occurs by the binding of the first component, the C1 complex, to an immune complex. C1 can also be activated by bacterial lipo-polysaccharides, DNA, RNA viral membranes etc., The importance of such activation is, however, not clear. C1 complex is made up of three subcomponents, C1q, C1r and C1s. C1q interacts with Clq binding sites of the C_H2 domain of the immunoglobulin in the immune complex. The C1q molecule has 18 polypeptide chains (six A, six B and six C), each of approximately 80 amino acids.

The amino terminal end of each chain has collagen like (-Glyx -Xaa - Yaa-) repeat sequences. This is followed by a C terminal portion of about 136 amino acids, which are non collagen like. It has the appearance of a bunch of tulips and the globular head region interacts with the antibody molecules in the immune complex. **(Figure 8.1)** Radial arms of C1q molecule interact with a tetrameric complex of two C1r and two C1s molecules.

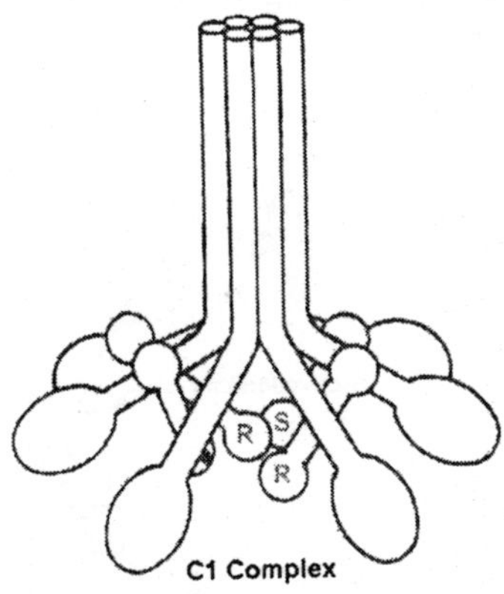

Figure 8.1 The subcomponents of C1 designated C1qrs.

Each Cl molecule must bind by its Clq globular heads to at least two Fc sites for a stable Cl - antibody interaction. Pentamer IgM bound to an antigen has at least 3 exposed sites for Clq. C1q has a weak binding with monomeric IgG as IgG has only single Clq binding site in the C_H2 domain of the Ig Fc region. Binding of the multiple heads to the many Fc sites of aggregated IgG in an immune complex increases strength of binding. In man, IgG4 is inactive in complement activation.

Clr and Cls each has a single polypeptide chain (85 KDa), which on activations has serine protease activity. Two molecules of Clr and C1s bind to one C1q subunit to form a multimeric complex. The binding of the C1q "head" to an Ig causes co-operational changes, which result in proteolytic cleavage of Clr, which has serine esterase activity. Clr then cleaves the associated Cls molecule to active Cls.

The next important molecule in the cascade is C4, a 205 Kd heterotrimeric complex consisting of polypeptide chains termed α, β and γ. Cls hydrolyzes a small fragment (C4a) from the amino terminus of the C4, exposing a binding site on the larger fragment (C4b). One Cls molecule can generate multiple C4b molecules. As a result of this proteolytic step, the thioester bond in the α chain of the C4b fragment becomes susceptible to nucleophilic attack by amine and hydroxyl groups and makes the molecule highly unstable. This chemically reactive form of the molecule is known as metastable C4b. Most of the C4b thioester bonds rapidly react with water molecules, generating a short-lived inactive intermediate, iC4b. However, some C4b thioester bonds form covalent amide or ester bonds with cell surface molecules. There are two isoforms of C4 i.e. C4A and C4B.

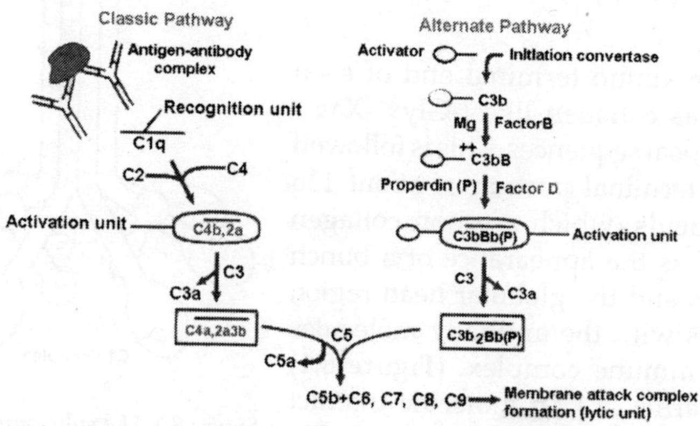

Figure 8.2 The classical and alternative pathways of the complement system

C2 is the third soluble component of the classical pathway and along with C4b is involved in forming the C3 convertase. It is a 102 Kd single chain polypeptides that binds to cell-surface bound C4b. C4b interacts with the N terminal domain of proenzyme C2 in a Mg++ dependent fashion and splits it at one point to yield the non catalytic C2b (39,000 mol wt). The major part of the molecule (C2a) stays bound to C4b at the cell surface and together they form the classical pathway C3 convertase (**Figure 8.2**)

Role of C3 in Classical Pathway

C3 holds a very important position in the complement system as the classical, lectin and alternate pathways converge at this point. C3 is a 195 Kd disulphide linked heterodimeric glycoprotein composed of α and β polypeptide chains. In the classical pathway, C3 is cleaved by C4bC2a which results in the removal of a 3Kd fragment, C3a, from the chain, and leaves a molecule, C3b. Removal of C3a causes conformational changes in the C3b portion and results in exposure of an internal thioester bond which is inaccessible in native C3. Binding of C3b takes place only in the immediate vicinity of C1 complex or of the C3 convertase enzyme. If no binding site is available, the reactive acyl group is hydrolysed by the aqueous environment and inactivated.

The association of C4 and C2 is unstable and C2, which is labile, can be lost from the covalantly bound C4 resulting in loss of C3 convertase activity. C4 can, however, react with new C2 molecules and form fresh C3 convertase enzyme.

The final event in the classical pathway is formation of C5 convertase. C3b interacts with already formed C3 convertase C4b2a, to generate the C5 convertase or C5 cleaving enzyme, C4b2a3b. Interaction of C3 with C 4b2a is essential to alter the substrate specificity of the active site on C2, to enable the enzyme to activate C5 (**Figure 8.3**).

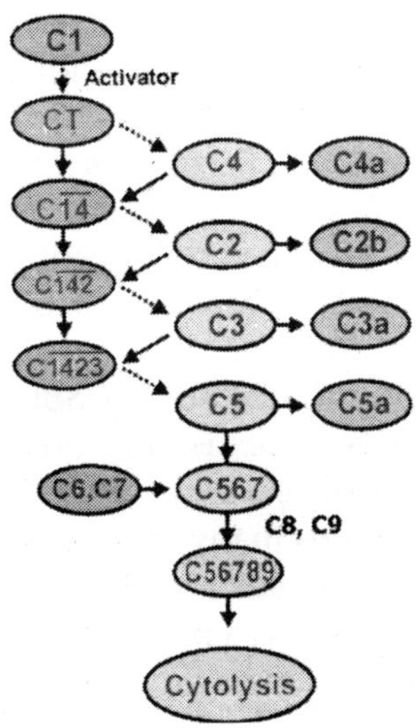

Figure 8. 3 The classical pathway of complement activation.

The Alternate Pathway

The alternate pathway differs from classical pathway in that its activation occurs in the absence of antibodies and both soluble and membrane bound forms of C3 convertases are generated.

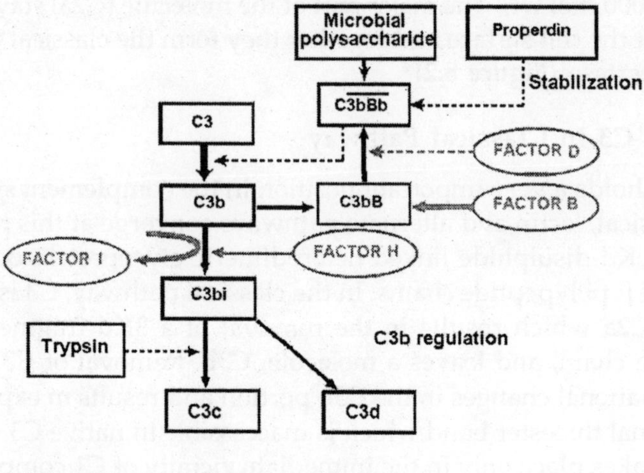

Figure 8. 4 Alternative complement pathway

The alternate pathway C3 is continuously activated at a low rate by serum proteases, which convert C3 to C3b. A continuous, slow spontaneous hydrolysis of internal thioester bonds within circulating C3 molecules occur without the loss of C3a fragment. The C3b that is generated, forms C3 convertase by binding to factor B in the presence of factor D. Factor B is a single chain 93 kD protein that is homologous to C2 of the classical pathway. After it forms the complexes with C3b or C3 it becomes susceptible to proteolysis by factor D, a 24 kD serine protease. Cleavage of bound factor B releases a 33 KD fragments (Ba) and a 63 kD fragment Bb is left attached to C3b or C3. The resulting C3bBb or C3Bb is the alternative pathway C3 convertase. This complex is very unstable and has to be stabilized by binding of properdin, a 220 KD protein. The cell surface bound C3bBb complex generates many more C3b molecules, which in turn are deposited on the same surface. Thus, C3b is both a component of C3 convertase enzyme and a product generated by C3 convertase allowing a positive feedback amplification of the whole process (**Figure 8.4**).

Lectin pathway

In the lectin pathway MBL binds to carbohydrate residues on the surface of the activating bacteria or other substances and undergoes conformational change, analogous to C1q in the classical pathway. MASP1 becomes active and cleaves C4 and C2 to form C4b2a, whereas MASP2 appears to be able to cleave C3 directly. However, the exact details of the lectin pathway are not known.

Membrane Attack Complex

The late phase of the complement cascade begins when C5 is bound and then cleaved by either the alternative or classic pathway convertase into C5a and C5b. C5a, a small 11 kD fragment, is released and has a number of biological effects. C5b, a 80 kD two-chain structure, does not form a covalent bond with the surface of its target. C5b is rapidly inactivated unless bound to the next component in the cascade, C6, which stabilizes it. The C5b6 complex can bind C7, the third protein involved in membrane attack. The C5b67 complex is strongly hydrophobic and interacts with nearby membrane lipids and inserts into the lipid bilayer of cell membranes. In that location one C5b67 complex can accept one molecule of C8 and multiple molecules of C9 to form the membrane attack complex (MAC). This complex is a cylindrical transmembrane channel, with a hydrophobic outer surface, which associate with the membrane lipid of the bilayer and a hydrophilic core, through which the small ions and water can pass. The ionic enrichment of the extra cellular fluid then communicates with that inside the membrane and the cell fails to maintain its osmotic and chemical equilibrium. Water enters the cell because of the high internal osmotic pressure and the cell swells and bursts.

Regulation of complement system

Under normal conditions, host cells are resistant to the effects of complement due to the action of many regulatory proteins. In conditions where antigen-antibody complexes persist, the complement activation is continuous and the number of proteolytic fragments produced overwhelms the normal regulatory mechanisms.

Complement regulation takes place at many levels in the complement cascade and is mediated by membrane bound proteins, as well as soluble proteins in circulation (**Table 8.2**). These regulatory proteins protect host cells from potential damage caused by activation of complement during host defence. A general mechanism of regulation is the very labile nature of some of the proteins, which undergo spontaneous inactivation if they are not stabilized by other proteins.

Table 8.2 Biological Function of Complement Regulatory Protein

Protein	Molecules Size (Kd) characteristics	Serum Concentration (ug/ml) cellular distribution	Pathway affected	Function
C1 inhibitor (C1 INH)	104	240	Classical	Serine protease inhibitor covalently binds to C1r and C1s and dissociates C1r2S2 from C1q
C4bp	550 8 SCR, S	250	Classical and lectin	Block formation of C3 convertase (C4b2a), by binding C4b, cofactor for Factor I mediated cleavage of C4b
Factor H	150 20 SCRs	300 - 450	Alternate	Blocks formation of C3 convertase by binding C3b; cofactor for Factor I mediated cleavage of C3b.
DAF	70	Widely distributed	Classical and alternate	Accelerates dissociation of C4b2a and C3bBf
Factor-1	88	35	Classical alternate and lectin	Serine protease, cleaves and inactivates C4b and C3b using C4bp factor H, CR1 or MCP as cofactors
S protein	83 (Vitronectin)	500	Terminal	Binds C5b67 and prevents its membrane insertion.
Homologous restriction factor (HRF)	65 Phosphati dylinositol linkage	Most blood cells, epithelial cells	Terminal	Bind to C5b678 on autologous cells, blocking binding of C3.

Protein	Molecules Size (Kd) molecular characteristics	Serum Concentration (ug/ml) or cellular distribution	Pathway affected	Function
CD 59 (Membrane inhibitor of reactive lysis (MIRL)	18 phosphatidylinositol linkage	Erythrocytes, Lymphocytes, Monocytes, Neutrophils, Platelets, Endothelial cells, Epithelial cells	Terminal	Bind to C5b678 on autologous cells, blocking binding of C3.
SP - 40, 44 Heterodimer N	80	50	Terminal	Modulate MAC formation
Anaphylatoxin inactivator Carboxy peptidase N	310	35	Effector	Carboxypeptide N removal of terminal arginine residues; inactivates the anaphylatoxins

SCR = Short Consensus Repeat; MAC = Membrane Attack Complex.

Regulation of the classical pathway

An inhibitor of C1, a glycoprotein termed C1 esterase inhibitor, can complex with Clr_2s_2 causing it to dissociate from Clq, and prevent initiation of the cascade. The C3b generated by C3 convertase of the classical, alternate and lectin pathways has the capability to bind to near by cells. These 'bystander' cells are opsonized and taken up by phagocytes bearing C3b receptors or may be lysed by formation of membrane attack complex. However, normal cells are protected from damage because C3b undergoes rapid spontaneous hydrolysis by the time it has diffused 40 nm from the C3 convertase. The potential destruction of normal cells by C3b is limited also by a group of related regulatory proteins. These contain repeating amino acid sequence of approximately 60 residues termed short consensus repeats (SCRδ). All these proteins are encoded at a single location on chromosome 1 in humans, known as regulators of complement activation (RCA) gene cluster.

In the classical and lectin pathways soluble C4b binding protein (C4bBP) and two membrane bound proteins, complement receptor type 1 (CR1) and membrane cofactor protein (MCP) act to prevent assembly of C3 convertase. These proteins prevent association of C4b and C2a by binding to C4b. Factor I cleaves the bound C4b to bound C4b and soluble C4c.

Regulation of Alternative Pathway

As in the classical pathway, regulation takes place primarily by inhibiting C3 convertase activity. Factor H encoded by a member of the RCA gene cluster, MCP and CR1 prevent the binding of C3b to factor B. Decay accelerating factor (DAF), factor H, C4b BP and complement receptor type 2 (CR2) accelerate the dissociation of C3bBb, the C3 convertase enzyme, promoting the release of C3b from the complex. It is homologous to the control of classical pathway C4b2a when it is bound to cell membranes. The inhibition of C3b is associated with Factor I which catabolizes C3b to C3c and C3d. Bacterial cell surface act as a protected site for C3b, since factor B has a higher affinity for C3b than does factor H at these sites.

Regulation of membrane attack complex

Three proteins CD59, HRF and S proteins regulate the terminal pathway of complement. The presence of these proteins prevent the formation of the membrane attack complex on the surface of autologous cells.

Decay-accelerating factor (DAF) is a single chain membrane protein (MW 70,000) that is a potent accelerator of C3 convertase decay, but unlike CR1 and CR3 it has no factor I cofactor activity. Functionally, the protein acts to limit membrane damage that may occur if complement is activated at the cell surface. C8-binding protein, also known as homologous restriction factor (HRF), acts to prevent completion and membrane insertion of the MAC. This membrane protein therefore acts to prevent cell lysis at yet another step in the complement cascade. It is called homologous restriction factor because it recognizes C8 and C9 of the same species better than C8 from other species.

CD59 is another regulatory protein that prevents assembly of the complete C5b-9 complex on target cell and so prevents complement-mediated lysis. Interestingly DAF, HRF and CD59 are each bound to the cell surface by a transmembrane domain within the amino acid backbone of the protein. This phosphoniositide linkage is reported to give the protein far greater lateral mobility within the cell membrane, increasing its ability to intercept damage causing complement complexes.

The complement regulatory proteins are species specific and so do not protect xenograft, a graft from one species transplanted into another species. For example in a porcine to man liver graft, porcine CD59 and HRF present in the graft will not protect the graft from the effects of human derived complement. On normal red blood cell, CD59 and HRF proteins are linked to the cell surface via special GPI (glycosyl phosphatidylinositol) linkages.

Biological Functions of Complement Proteins

The functions of the complement system include the cell lysis mediated by MAC and various biologic effects of the proteolytic fragments of complement. Complement activation occurs on cell surfaces leading to the final insertion of MAC into the bilipid layers, causing osmotic lysis of cells. The proteolytic fragments generated during complement activation either remain bound to the cell membrane at the site of complement activation or are released into the fluid phase. Complement is involved in killing of invading micro-organisms directly and enhancing ingestion and killing by phagocytic cells. All the three pathways are involved in this process. However, alternate pathway is the most efficient in causing killing and opsonization.

Individuals that are missing one of the alternative pathway proteins, late components (C3 - C9) or one of the control proteins (H or I) tend to get severe infections with pyogenic organisms. They are particularly susceptible to infection by gram-negative bacteria, particularly *N. meningitides*. Deficiencies of the classical pathway components (C1, C4 , C2) are also associated with increased risk of infection, though not as frequently as with the alternative pathway or late component deficiencies. Bacteria that cause recurrent infections in these patients include, but are not limited to *Streptococcus pneumoniae, Haemophilus influenzae,* and *Staphylococcus aureus*. Complement deficiencies are relatively rare (1 in 10,000 people), and are not always associated with disease, since many complement - deficient individuals are diagnosed only after a family member has been identified.

Complement also has the ability to neutralize viruses. C1 can bind directly to the surface of several viruses, including the lentiviruses, HIV-1, and HTLV-1. MBL also binds and is activated by of HIV-1, HIV-2, and influenza virus. Antibodies against the viruses mediate further binding and activation of the classical pathway while repeating subunits on the viral capsid or membrane surfaces activate the alternative pathway. The binding of complement proteins not only leads to opsonization of the virus and lysis of the virion, but it also interferes with ability to interact with the membrane of its target cells, and thus blocks its entry into the cell.

Complement interaction with viruses has proved to be double-edged sword. While complement clearly plays a role in defence against viral infection, many viruses have evolved mechanism that take advantage of complement proteins, both for controlling complement deposition on their surface and for gaining entry in target cells. Paramyxoviruses (measles virus) use MCP as a receptor for entry into host cells, while viruses of the picorna virus family use DAF to infect epithelial cells. The Epstein Barr virus uses CR2 to gain entry into mononuclear cells. Viruses capture complement control proteins when the virions bud from the cell membranes. HIV-1, HTLV-1, simian immunodeficiency virus (SIV), and cytomegalovirus (CMV) capture DAF, MCP, and CD59 on their surface and are thus protected from complement-mediated lysis in the same way that host cells are protected. Several viruses also produce viral proteins that mimic complement inhibitor function. The herpes viruses make several proteins that have DAF and / or MCP - like activities, or that block C5b-9 formation, and vaccinia virus produces vaccinia control protein (VCP) that contains 4 CCP domains, binds to C3b and C4b and has both decay - accelerating and cofactor activities.

Complement plays a protective role during growth of the fetus in the uterus. The tissues of the female reproductive tract and the gametes have abundant regulatory proteins. MCP DAF, and CD59 also protect spermatozoa from complement attack. The uterus in response to estrogen, secretes C3 and this secretion is blocked by progesterone. MCP and DAF are located variably on fetal tissues, while CD59 is present on all tissues throughout pregnancy. Tissues at the maternal-fetal interface are at particular risk to complement regulatory proteins.

Another important function of complement is in the clearance of immune complexes, (**Figure 8.5**). Antibodies bind to their appropriate antigens and cross linking between the molecules increase the size of the complex until they become insoluble. This is detrimental, as large insoluble complexes accumulate in tissues such as the skin and kidneys where they cause inflammation and damage to the surrounding cells. Once several antibodies bind to an antigen and become cross-linked, C1 activation occurs and C4b and C3b are deposited on the molecules of the complex. The incorporation of the relatively large complement proteins interferes with further binding of immunoglobulins, and the

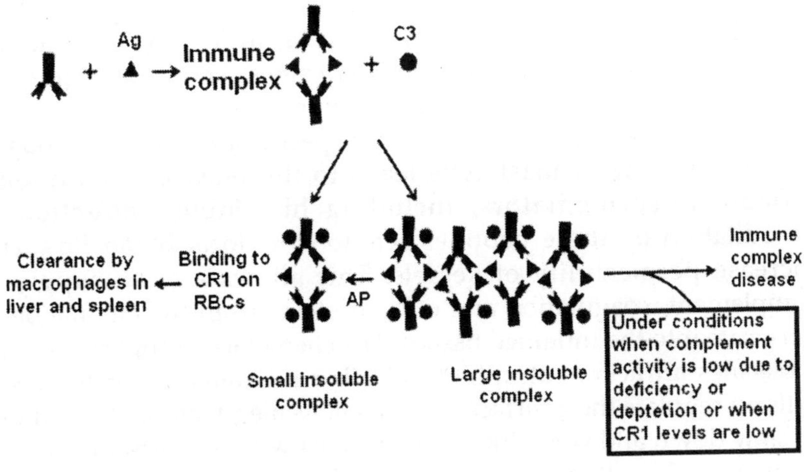

Figure 8.5 Role of complement in clearance of immune complexes

complex can now bind to C3 receptors and be removed from the circulation. If the complex becomes very large in spite of C1 activation, or if the immunoglobulin is of a class that does not bind C1, alternative pathway activation can occur with more C3b deposition. This interferes with the bonding forces that keep the complex together and it starts breaking up into smaller molecules that can be cleared by C3 receptors on the cells of reticulo-endothelial system.

One of the most important mechanisms of immune complex clearance is by the CR1 receptors on erythrocytes. Although there are more CR1 molecules per cell on leukocytes than on erythrocytes, there are far more erythrocytes in the circulation and the major part of total available CR1 is on these cells. Small soluble immune complexes with C3b on their surfaces become bound to the CR1 and are carried to the liver and spleen. At these sites, the complex is transferred from the erythrocyte CR1 to macrophage CR3 and Fc receptors and taken into the phagocytes and destroyed. If there is a deficiency in the classical pathway components, large insoluble immune complexes can form resulting in immune-complex mediated disease.

In addition to clearing immune complexes, complement aids in the clearance of dead tissues and foreign substances. Subcellular membranes from mitochondria, endoplasmic reticulum or other organelles directly activate both classical and alternative pathways, rendering them recognizable by phagocytic cells that have C3 receptors. Unlike cells killed by infection, ischemia, or other means, apoptotic cells do not activate complement.

The anaphylatoxin fragments C3a, and C5a, have many physiological effects at the inflammatory site as shown in **Figure 8.6**. C3a and C5a induce smooth muscle contraction of ileum, bronchus, uterus and blood vessels and in most species studied. Interaction of C3a and C5a with mast cells leads to the release of additional inflammatory mediators, including histamine. Induction of vasodilation by these peptides acts to slow local blood flow and increase permeability of vessels. This allows a fresh supply of complement, coagulation and other proteins from the circulation to permeate into the inflamed tissues. The chemotactic property of these molecules attracts neutrophils and other inflammatory cells. These cells up regulate their surface integrin molecules, have an opportunity to slow down and come into better contact with the vessel wall, bind by way of the adhesion molecules, and leave the circulation for the site of inflammation. C5a also interacts with monocytes and hepatocytes to cause the release of cytokines and other mediators. IL-6 and C5a induce the acute phase response. Serum carboxypeptidase N (SCP-N, chemotactic factor inactivator, CFI) rapidly cleaves the C-terminal arginine from C3a, C4a and C5a. The des-Arg form of the anaphylatoxins is unable to bind productively to their receptors, although $C5a_{des\text{-}Arg}$ retains about 2% of its previous chemotactic activity.

C3d, covalently bound to antigen with immune complexes is taken up by CR2 receptors in cells of the dendritic lineage and are retained for long periods. This forms a continuous source of antigen that stimulates B-cell responses and the memory response. The immune response to a T- cell dependent antigen is impaired in complement - deficient animals, and the isotype switch from IgM to IgG does not occur unless the system is driven by very large amount of antigen. Studies using strains of mice deficient in C3 or C4 have proved that the enhancing effect of complement on the acquired immune response is dependent on the amount of the components

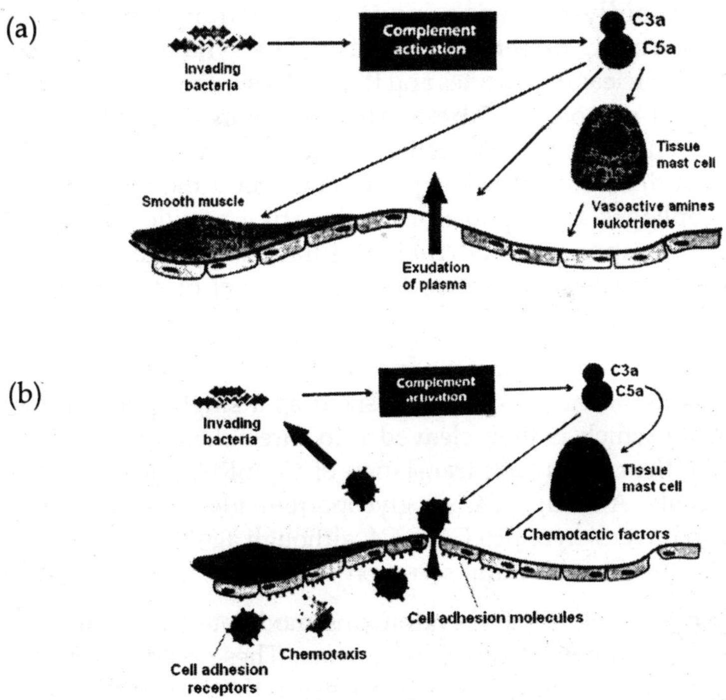

Figure 8.6 Role of complement in inflammation: (a) anaphylatoxins and (b) chemotaxis

present. An animal that is heterozygous for a complement deficiency will have a response that is intermediate between a normal animal and one that has a complete deficiency.

The complement - dependent enhancement of the immune response is mediated by the binding of CR2 to C3dg on the antigen, followed by cross linking of the antigen through surface IgM on the B-cell surface. CR1 is often found with CR2 on the B cell and may bind C3b -bearing immune complexes that can then be processed to C3dg by the action of factor I, and transferred to CR2. Human CR2 is also found in a complex with CD19, CD81, and Leu-13 so that transmembrane signals are transmitted via phosphorylation of CD19 when CR2 and mIgM are cross linked by the C3dg. Binding of C3dg to the antigen also appears to alter the enzymatic digestion of the antigen as it is processed by antigen presenting cells. Antigen bound C3d appears to have an adjuvant effect as well.

Biosynthesis of Complement Proteins

Complement proteins are acute phase molecules that are synthesized rapidly following injury or infection. Most of the components normally present in the plasma are synthesized by the liver and mononuclear phagocytes and the rate of synthesis can rapidly be increased in response to stress. The membrane-bound control proteins are synthesized by the cell on which they are expressed. It has also been demonstrated that peripheral blood monocytes and tissue macrophages are capable of synthesizing a wide variety of complement components and their regulatory proteins. This is important at extra vascular sites, where the level of complement protein is usually low.

Little is known regarding the biosynthesis and secretion of the majority of components. C4 is synthesized as a single polypeptide chain (pro-C4) which is then cleaved into three chains before it is secreted from the cell. *In vitro* translation of C4 mRNA produces the pro-C4 molecule. Although C4 is a glycoportein, glycosylation is not linked to the conversion of pro-C4 to C4, although agents which block glycosylation result in blocking secretion of C4 molecule.

Hormones, cytokines and endotoxins modulate the amount of complement components present in plasma. These modulators are generally thought to be of significance for the biosynthesis of complement components by phagocytic cells.

Complement receptors

Many of the biological activities of the complement system depend on the binding of complement fragments to complement receptors. These are expressed on a variety of cell types. Some of the complement receptors also may have a role in regulating complement by binding biologically-active complement components and breaking them into inactive compounds. The receptors, their ligands and cellular distribution are given in **Table 8.3**.

The Genetics of the Complement Proteins

Structure and function of complement proteins and genes

Some complement components share structural characteristics and have similar functions. These proteins may be grouped into families, which may have evolved from common ancestor genes. For example, four genes, C4A, C4B, Bf and C2 map to the short arm of chromosome 6 within the major histocompatibility complex (MHC). These genes are collectively known as class III MHC genes. This region is located downstream of the class II MHC genes (DR, DQ, DP) and upstream of the class I MHC genes (B and C). C2 and Factor B of the alternate pathway share extensive protein sequence homology. Both participate in the formation of macromolecules, which activate C3, and both are Mg_2^+ -dependent serine esterases. The high degree of homology and close functional similarly between the C4A and C4B genes and between the Bf and C2 genes, suggests that duplication of ancestral genes may have resulted in these two sets of genes.

The class III genes on chromosome 6 span a 120-kilobase (kb) segment approximately 390 kb telomeric to HLA-DR and 600 kb centromeric to HLA-B. Specific variants of class III genes have been recognized to segregate more commonly with specific MHC genes. For example, the most common deletion within the C4A Q0, B1 allotypes. Such combinations of alleles have been termed "extended haplotypes". The HLA-A1, B8, DR3, C4AQ0, B1 haplotype is associated with several autoimmune diseases such as diabetes, membranoproliferative glomerulonephritis, and systemic lupus erythematosus. The conversion of a C4B gene to a C4A gene (producing an apparent C4B "null" allele) is found commonly on the extended haplotype HLA - B44, DR6 and on the haplotype HLA - B35, DR1, C4A3, 2, BQ0.

Two genetic mechanisms produce C2 deficiency. The common

Table 8.3 Complement receptors, their major ligands and their cellular activity

Receptor	Major ligands	Activity	Cellular distribution
CR1 (CD35)	C3b, C4b	Blocks formation of C3 convertase; binds immune complexes to cells	Erythrocytes, neutrophils, mono-cytes, macrophages, eosinophils, follicular dendritic cells, B cells, some T cells
CR2 (CD21)	C3d, C3dg*, iC3b	Part of B- cell coreceptor; binds Epstein –Barr virus	B cells, follicular dendritic cells, some T cells
CR3 (CD11b/ 18) CR4 (CD11c/18)	iC3b	Bind cell –adhesion molecules on neutrophils, facilitating their extravasation; bind immune complexes, enhancing their phagocytosis	Monocytes, macrophages, neutrophils, natural killer cells, some T cells
C3a / C4a receptor	C3a, C4a	Induces deregulation of mast cells and basophils	Mast cells, basophils, granulocytes
C5a receptor	C5a	Induces deregulation of mast cells and basophils	Mast cells, basophils, granulocytes, monocytes, macrophages, platelets, endothelial cells

* Cleavage of C3dg by serum proteases generates Ced and C3g.

mechanism that results in a C2 null allele is a 28-base pair (bp) deletion within the C2 gene. C2 deficiency has been estimated to have prevalence between 1:10,000 and 1:30,000 and the C2 null allele is strongly linked with the haplotype HLA A25, B18, DR2, BfS, C4A4, B2. This particular 28-bp deletion has been reported to be common in Caucasian SLE patients and absent in African-American SLE patients.

C1 and C8 are unique among the complement components, in that these proteins are encoded by more than one gene. The genes for C1r and C1s are present together on chromosome 12 and the serum proteins function as serine proteases as part of the macromolecule, C1. The gene for C1q, which encodes the third molecule that constitutes C1, is located on chromosome 1. C8 has three chains (α, β and η) and closely linked genes on chromosome 1 code for the α chain and β chains, whereas the γ chain is encoded by a gene on chromosome 9.

C5 maps to chromosome 9. The C5 protein shares both structure and function with C3 and C4. The genes for C6 and C7, both map to chromosome and encode proteins that are structurally related.

On chromosome 1, the group of regulators of complement (RCA) genes encode factor H, C4b-binding protein (C4bp), decay-accelerating factor (DAF), membrane cofactor protein (MCP), and the complement receptors CR1 and CR2. Each of these inhibitors contains varying numbers of the same short consensus repeats (SCR) of 60 to 70 amino acids, which indicate the probability of a common ancestor. The SCRs are also found on other complement C components, which share the ability to bind to fragments of C3.

The molecular basis of complement polymorphism

Employing protein electrophoresis methods combined with specific antibody precipitation or other functionally specific detection systems, complement phenotypes and their variants have been described. Pedigree analysis and techniques of molecular biology have permitted the chromosomal assignment for these C genes.

C4 genes, C4A and C4B (each encoding, respectively, C4A and C4B isotypes or classes), exhibit the most extensive polymorphism among complement proteins. The isotype determinants for C4A and C4B are encoded by 6 base pairs in exon 26 of the C4 genes. Other functional sites within the C4 genes have been determined for the C1 cleavage site, the C3b –binding site, the thioester site, and the C5b –

binding site. At least 14 variants of C4A and 17 variants of C4B have been reported. Rare C4 variants display "mosaic" characteristics in which features of both C4 isotypes can be detected.

The molecular basis of non-functional and "null" variant complement protein has been described. Null genes for C4 have been recognized surprisingly frequently and arise from several different mechanisms. The most common cause of a C4 null allele is a large 28-kb deletion of the 5′ end of the C4 gene. C4A null allele may be due to the insertion of thymidine –cytosine, a mutation that produces a frame shift and stop codon within this particular allele. A third mechanism producing C4 null alleles is due to the conversion of one isotype to the other usually C4B to C4A. This mutation results in an "apparently null" allele when phenotyping is assessed by protein electrophoresis. Although two C4 alleles are actually present they encode only one C4 isotype, leading to the interpretation that the paired C4 isotype gene is absent. Some C4 null alleles occur from as yet unidentified mutations.

CHAPTER 9

Immunodeficiencies

Immunodeficiencies, as the name suggests are a group of diseases characterized by a defective immune system and may either be inherited (termed primary immunodeficiency) or acquired (termed secondary immunodeficiency). **Acquired Immunodeficiencies Syndrome** (AIDS) is the characteristic example of a secondary immunodeficiency. This chapter will, however, restrict the discussion to inherited primary immunodeficiencies.

The genetic immunodeficiency diseases were first recognized in the twentieth century, when the newly discovered antibiotics failed to control infections in some infants. In 1952, Bruton made the first definite report of an 8-year-old boy who had recurrent episodes of sepsis and was found to have agammaglobulinemia; the condition has named after Bruton. In 1965, DiGeorge described the association of infections with absence of the thymus gland in three children. These children had defects of T cells, which could be corrected with thymus transplants. This suggested that lymphocyte development in the thymus was essential for protection against infections. Over 50 primary immunodeficiencies have been described and the progress in the field of genetic immunodeficiencies has run parallel to that of immunology itself, contributing in great measure to the understanding of the immune system.

The lymphocyte was the least important white blood cell till the middle of the twentieth century and its functions were ill understood. The findings of DiGeorge marked the birth of "cellular Immunology" and was followed by rapid unravelling of the immune system using murine models in which it was possible to do surgical manipulations such as removal of the thymus at birth and cell transfer experiments.

Congenital abnormalities of almost every component of the immune system have since been described and the defects can be classified according to the subdivisions of the immune system as defects of:

- B lymphocyte (antibody) system
- T lymphocyte (cellular immune) system

- Phagocytic system
- Complement system

This division provides an easy way to classify the congenital immunodeficiencies and they will be discussed in the same manner in this chapter. However, for the sake of completion, the official classification of International Union of Immunological Societies is given in **Table 9.1**. This is a more complex classification, which takes into account all the interactions between B and T cells.

Table 9.1 WHO classification of primary and secondary immunodeficiencies

Groups	Inheritance*
A Predominantly antibody deficiencies	
1 X-linked agammaglobulinemia	XL
2 Hyper-IgM syndrome	
X-linked	XL
Other	AR, unknown
3 Ig heavy-chain gene deletions	AR
4 k-Chain deficiency	AR
5 Selective deficiency of IgG subclasses (with or without IgA deficiency)	AR
6 Antibody deficiency with normal IgG	Unknown
7 Common variable immunodeficiency	Various: AR, AD or unknown
8 IgA deficiency	Various: AR, unknown
9 Transient hypogammaglobulinemia of infancy	Unknown
B Combined immunodeficiencies	
10 Severe combined immunodeficiency (SCID)	
X-linked	XL
Autosomal recessive	AR
11 Adenosine deaminase (ADA) deficiency	AR
12 Purine nucleoside phosphorylase (PNP) deficiency	AR
13 MHC class II deficiency	AR
14 Reticular dysgenesis	AR

Groups	Inheritance*
15 CD3g or CD3e deficiency	AR
16 CD8 deficiency	AR
C Other Well-Defined immunodeficiency syndromes	
17 Wiskott-Aldrich syndrome	XL
18 Ataxia-telangiectasia	AR
19 DiGeorge anomaly	Unknown
D Other primary immunodeficiency diseases	
20 Primary CD4 deficiency	
21 Primary CD7 deficiency	
22 IL-2 deficiency	
23 Multiple cytokine deficiency	
24 Signal transduction deficiency	
E Defects of Phagocytic Function	
25 Chronic granulomatous disease (CGD)	
X-linked CGD (deficiency of 91 kD binding chain of cytochrome b,	XL
Atutosomal recessive CGD	
p22 phox	
p47 phox	AR
p67 phox	
26 Leukocyte adhesion defect (deficiency of β chain (CD18) of LFA-1, Mac-1, p 150, 95)	AR
27 Leukocyte adhesion defect 2 (failure to convert GDP mannose to fucose)	
28 Neutrophil G6PD deficiency	XL
29 Myeloperoxidase deficiency	AR
30 Secondary granuie deficiency	AR
31 Shwachman syndrome	AR
F Compiement deficiencies	
32 C1q deficiency	AR
33 C1r deficiency	AR
34 C4 deficiency	AR
35 C2 deficiency	AR
36 C3 deficiency	AR
37 C5 deficiency	AR

Groups	Inheritance*
F Complement deficiencies continued	
38 C6 deficiency	AR
39 C7	AR
40 C8a + C8g deficiency	AR
41 C8b deficiency	AR
42 C9 deficiency	AR
43 C1 inhibitor deficiency	AD
44 Factor I deficiency	AR
45 Factor H deficiency	AR
46 Factor D deficiency	
47 Properdin deficiency	XL
G Immunodeficiency Associated with or secondary to other diseases	
Chromosomal instability or defective repair	
48 Bloom syndrome	
49 Fanconi anemia	
50 ICF syndrome	
51 Nijmegen breakage syndrome	
52 Seckel syndrome	
53 Xeroderma syndrome	
Chromosomal defects	
54 Down syndrome	
55 Turner syndrome	
56 Chromosome 18 rings and deletions	
Skeletal abnormalities	
57 Short-limbed skeletal dysplasia	
58 Cartilage-hair hypoplasia	
Immunodeficiency with generalized growth retardation	
59 Schmike immuno-osseous dysplasia	
60 Immunodeficiency with absent thumbs	
61 Dubowitz syndrome	
62 Growth retardation, facial anomalies, and immunodeficiency	
63 Progeria (Hutchinson-Gilford syndrome)	
Immunodeficiency with dermatologic defects	
64 Partial alpinism	

Groups	Inheritance*
65 Dyskeratosis congenital	
66 Netherton syndrome	
67 Acrodermatitis enteropathica	
68 Anhidrotic ectodermal dysplasia	
69 Papillon-lefevre syndrome	
Hereditary metabolic defects	
70 Transcobalamin 2 deficiency	
71 Methylmalonic academia	
72 Type I hereditary orotic aciduria	
73 Biotin-dependent carboxylase deficiency	
74 Mannosidosis	
75 Glycogen storage disease, type 1b	
76 Chediak-higashi syndrome	
77 Hypercatabolism of immunoglobulin	
78 Intestinal lymphangiectasia	
Other	
79 Hyper -IgE syndrome	
80 Chronic mucocutaneous candidiasis	
81 Hereditary or congenital hyposplenia or asplenia	
82 Lvermark syndrome	

Abbreviations: AD= autosomal dominant; AR= autosomal recessive; GDP= guanosine diphosphate; G6DP= glucose 6 phosphate dehydrog leukocyte function associated antigen; Mac= macrophage; phox= phagocyte oxidase; XL= X-linked.

It should be emphasized that the primary immunodeficiencies are a rare group of diseases with an incidence of approximately one per 10,000 of population (excluding asymptomatic IgA deficiency). In the majority of cases they are recognized in the first 2 years of life. These children usually present to the physician with recurrent or overwhelming infections. The pattern of infection, the type of organism responsible and the presence of associated findings like the absence of a thymic shadow on chest X-ray, gives an indication of the type of defect one is dealing with. (**Table 9.2**)

Table 9.2 Predominant infections and suggestive clinical findings which suggest nature of immune defect.

Predominant infections	Physical examination	Type of immunodeficiency
Repeated pyogenic infections with capsulated organisms like *Streptococcus pneumoniae, streptococcus pyogenes, Hemophilus influenzae* causing sinusitis, bronchitis, tonsillitis, otitis media, pneumonia	Atrophic peripheral lymph nodes and tonsils	B-lymphocyte deficiency
Predisposition to infections by intracellular pathogens like mycobacteria, *Salmonella, Listeria, Toxoplasma*, fungi and viruses; development of active infection after administration of a live virus vaccine		T-lymphocyte deficiency
Neonatal tetany with cardiac malformations, mongoloid facies, severe viral and mycotic infection	Lack of thymic shadow on x-ray	DiGeorge syndrome
Abscess-forming infection with lowgrade pathogens (*Staphylococcus epidermidis, Serratia marcescens, Aspergillus* sp).	Scars from previous bone, liver, or soft tissue abscesses, draining lymphadenitis; hepatosplenomegaly	Neutrophil deficiency
Repeated infection with *Neisseria* sp.		Complement deficiency
Persistent infection of the lungs, diffuse mucosal moniliasis, chronic diarrhea, and wasting early in life.	Growth retardation; other associated congenital abnormalities	Severe combined immunodeficiency

Primary Antibody Deficiencies

These are a group of conditions characterized by an impaired ability to produce antibodies. They are caused by mutations in genes involved in B cell differentiation or in the immunoglobulin (Ig) genes themselves. In some, the underlying defect is unclear. The typical patient will present with a history of recurrent upper respiratory or lung infections and have reduced serum levels of one or more of the Ig classes. The list of diseases in this group is given in **Table 9.3.**

The disorders of humoral immunity are the products of specific defects in B-cell differentiation. In brief, B cells are generated in fetal liver and adult bone marrow. Contact with cognate antigen (e.g. polysaccharide) activates the B cell and allows it to progress down the memory or plasma cell differentiation pathway. The response to protein antigens requires T cell help. In the germinal centres of lymph nodes, B cells can switch the heavy chain isotype and undergo somatic hypermutation, resulting in affinity maturation.

Table 9.3. Primary antibody deficiencies

X-linked agammaglobulinemia (Bruton tyrosine kinase (btk) deficiency, XLA)
Autosomal recessive agammaglobulinemia
μ heavy chain disease
Surrogate light chain (λ5/14.1) deficiency
Hyper-IgM syndrome (HIM)
X-linked (CD154 or CD40 ligand deficiency, X-HIM)
Autosomal recessive AID
IgA deficiency (IgAD)
Common variable immunodeficiency (CVID)
Selective IgG subclass deficiency
Constant domain deletions
Selective light chain deficiency
Transient hypogammaglobulinemia of infancy (THGI)
Antibody deficiency with normal serum immunoglobulin levels
VκA2 deficiency

X- linked Agammaglobulinemia (XLA)

XLA, which is also called Bruton agammaglobulinemia, is due to loss-of-function mutations in Bruton's tyrosine kinase (btk) gene on the X chromosome, which leads to a block in B-cell maturation. The affected males have absent or low serum immunoglobulin (Ig) levels, reduced or absent B cells in peripheral blood and lymphoid tissues and no germinal centres in lymph nodes. The T cell areas in the lymphoid tissues are normal as are T cell numbers in the peripheral blood. The number of heavy chain synthesizing pre-B cells in the bone marrow is normal implicating the genetic defect to the point of maturation of pre-B cells to surface μ positive B cells. The *btk* gene is located on the long arm of the X chromosome (Xq21.3-22) and codes for a protein tyrosine kinase of the subfamily of Src cytoplasmic protein-tyrosine kinases. This gene is normally expressed in all stages of B cell differentiation. Affected persons have normal numbers of pre-B cells. However, further differentiation is affected as the enzyme is essential for signalling events that are needed for light chain gene rearrangement.

Autosomal Recessive agammaglobulinemia

This condition is rare and occurs due to loss-of-function mutations in the components of the B cell receptor. Mutations in both alleles of the μ heavy chain region (μ_0) with complete B-cell deficiency and mutations in both alleles of the surrogate light chain receptor λ5/14.1 ($\lambda 5_0$) with reduced numbers of B-cells have been described. In both situations there is severe agammaglobulinemia and diagnosis requires mutation analysis.

Hyper-IgM syndrome (HIM)

This condition is characterized by markedly reduced levels of serum IgG, IgA and IgE with normal or increased levels of IgM and normal numbers of B cells. The heavy chain genes for γ, α and ε are normal but the isotype switch mechanism is defective; the genetic defect has been mapped to the short arm of the X chromosome. The molecular basis for this defect is a loss-of-function mutation in the CD154 gene, which codes for the ligand for CD40 (CD40L). CD 40L is expressed on activated T lymphocytes and interaction of CD40L with CD40 on B cells is essential for isotype switch. The gene for CD40L has been mapped to Xq26, where the hyper-IgM syndrome was previously mapped. However, autosomal recessive and autosomal dominant forms of the disease have also been reported.

IgA Deficiency

Selective IgA deficiency is numerically the most common of all the inherited immunodeficiencies with a reported frequency of 1 in 600 persons of European ancestry. It is inherited both as autosomal dominant and recessive traits and is characterised by very low serum IgA (usually < 5 mg/ml) with normal or elevated levels of IgM and IgG. B cells successfully rearrange heavy chains and have normal isotype switch with expression of membrane IgA; but the transition to IgA secreting plasma cells is blocked. The T cells are normal in number and function. The exact nature of the defect in this condition is not clear. Majority of individuals remain asymptomatic and deficiency is discovered incidentally. However, it is important to remember that theses individuals can produce IgG and IgE anti-IgA antibodies and hence are at high risk of adverse reactions following transfusions with blood products or preparations of intravenous immunoglobulin containing IgA. Individuals with high anti-IgA titers (>1:1000) are at risk of severe anaphylaxis. Individuals with IgA deficiency have a strong predisposition to develop a wide range of autoimmune diseases including systemic lupus erythematosus, chronic hepatitis etc.

Common Variable Immunodeficiency

This is an inherited autosomal disease in which patients usually present at variable times in late adolescence or early adult life. It is characterized by a broad deficiency of Ig isotypes with normal number of surface-Ig positive B-cell precursors in the blood. IgG and IgA levels in the serum are low and some IgG subclasses are more affected than others with the sequential order of involvement being IgG_4 > IgG_2 > IgG_1 >IgG_3. Some patients may also have low levels of IgM and IgE. The B cell areas in lymphoid tissues are usually hyper plastic but plasma cells are absent. It is not clear whether the defect is in the terminal differentiation of B cells or at the level of T cell helps to the B cells There is a higher risk for the development of a broad spectrum of autoimmune diseases and there is an unusually high incidence of lympho reticular and gastrointestinal malignancies. Family members have an unusually high incidence of IgA deficiency.

Selective IgG subclass deficiencies

Selective IgG subclass deficiency occurs often with normal total serum IgG levels. The most common deficiencies are of IgG_3 in adults and of IgG_2 in children. These defects are not due to deletions of the γ chain genes but probably some differentiation defect in the B cells themselves.

Ig heavy-chain deletions

Approximately 5-10% of Caucasians have deletions or duplications in chromosome 14q32 of the heavy chain constant region gene. Individuals with the homozygous deletions fail to make the corresponding isotypes while heterozygotes are unaffected.

Selective light chain deficiency

This is a very rare condition in which there is selective deficiency of κ or λ light chains.

Transient hypogammaglobulinemia of infancy

Infants suffer a physiological defect in their serum Ig at 4 to 6 months. This is the period when the maternally transmitted IgG wanes and endogenous production has not achieved normal values. During this period, child is prone to infections.

Selective antibody deficiency with normal Ig

This condition is characterised by failure to respond to some antigens, usually polysaccharide antigens, although there is no abnormality of other Ig or of the B cells. Most affected persons are normal with an occasional affected person presenting with recurrent sino-pulmonary infections.

Primary T-cell Immunodeficiencies

This group of conditions are caused by inherited defects of components of the immune system involved in T cell differentiation and function. Since the T-cell is central to the immune system, the defects in the T-cells not only affect T-cell functions but also the other arms of the immune response such as monocyte, and B cell. The affected children have severe, potentially lethal infections, lymphopenia and major abnormalities of T and B cell functions early in life. They can be classified into four groups according to the severity of and associated features:

- Severe combined immunodeficiencies (SCID),
- T-cell immunodeficiencies with detectable T cells
- T-cell immunodeficiencies with associated defects
- Miscellaneous T-cell immunodeficiencies.

Cellular immune deficiencies predispose to infections mainly by intracellular organisms. Many of these cause clinically severe disease, which is difficult to eradicate. These disorders are recognised by low numbers of T cells in the peripheral blood, reduced responsiveness to polyclonal T cell activators (PHA) and poor cutaneous delayed type hypersensitivity reactions to commonly present microbial antigens.

Severe Combined Immunodeficiencies (SCID)

This consists of a group of genetic disorders characterized by a marked defect in T-cell maturation, with or without an abnormal B-cell maturation. It is a rare condition with an overall frequency between 1: 50 000 and 1:100 000 live births. It is uniformly fatal if untreated. The entities, which make up this syndrome, are given in **Table 9.4**.

Table 9.4: Classification of SCID

Disease	Inheritance	Cells affected	Gene product
Reticular dysgenesis	AR	T,B, NK, leukocyte, platelet	?
Alymphocytosis	AR	T, B	RAG1, RAG2
Absence of T lymphocytes	XL	T, NK	γc chain
	AR	T, NK	JAK3
	AR	T	IL-7Rα
	AR	T +/- NK	?
ADA deficiency	AR	T, B, NK	ADA

AR=Autosomal recessive; XL= X-linked recessive

The clinical presentation is usually in the first few months of life with infections of the respiratory tract and gut. The affected infants do not have palpable lymph nodes or tonsils, have an absent thymic shadow in the chest skiagram with marked lymphocytopenia. Diagnosis can be confirmed by enumerating the T, B and NK cells by flow cytometry, assessing the functional capacity of lymphocytes or estimating the adenosine deaminase enzyme levels. Extremely low numbers of T cells is a common feature of various types of SCID.

Autosomal recessive SCID with absence of T and B-lymphocytes is found in 20% of all cases. This phenotype is characterized by defective V(D)J recombination of the T-cell receptor and Ig genes. Defective T and NK cell differentiation, characterizes X-linked SCID and accounts for 50% of all cases. In X-L SCID, B cell numbers are normal but are functionally abnormal due to lack of T-cell help. The molecular defect that has been identified in XL variety are mutations in the gene for the common γc chain of the IL-2 receptor (IL-2R), which is also shared, by the receptors for IL-4, IL-7, IL-9 and IL-15. The gene for γc chain is located at Xq12-13.1.

A T (-) B (+) NK (+) phenotype is found in approximately 10% cases; these have an autosomal recessive inheritance pattern. These patients have mutations in the Janus associated kinase (JAK) 3 gene.

The other major phenotype, which accounts for 20% of SCID cases, is ADA deficiency. Deficiency of the ADA enzyme, which is involved in the salvage pathway of purine metabolism, leads to abnormal accumulation of adenosine, which is toxic to T cells. The gene for ADA is located at 20q13-11.

T cell immunodeficiencies with detectable T cells (Combined immunodeficiency)

These diseases are characterized by recurrent infections, which may not be life-threatening and hence the disease may not manifest in the first few months of life as in SCID. Children develop repeated episodes of infection of chest and gastrointestinal tract, which are difficult to treat. Approximately 50% of them develop autoimmunity, most frequently of the blood cells and there is a high incidence of development of cancer. A large number of children also have allergic manifestations, such as eczema and asthma. The number of T cells in blood is often reduced and their function is impaired. A number of distinct syndromes have been described and in many of these, the molecular defect is yet to be characterized. **Table 9.5** provides a list of the disorders.

Major histocompatibility complex (MHC) class II deficiency (Bare lymphocyte syndrome)

This rare disease is characterized by a profound defect in the expression of HLA class II molecules. It has autosomal recessive inheritance. Since MHC class II expression is essential for thymic development of CD4+ cells, CD4+ T cell numbers are usually low although the total number of lymphocytes is normal. The antigen specific responses *in vivo* and *in vitro* are absent and the total immunoglobulin (Ig) levels and B cell functions are abnormal. The molecular basis of the defect is mutations in the transcription factor RFX that binds to the cis-acting X element of the MHC class II genes.

TAP1/TAP2 deficiencies

This condition is characterized by reduced expression of MHC

Table 9. 5 T cell immunodeficiencies with detectable T cells

Defect	Inheri-tance	Phenotype	Molecular defect
MHC class II deficiency			
(Bare lymphocyte syndrome)			Transactivator
A	AR	Defective expression of MHC class II	CIITA
B	AR	Defective expression of MHC class II	RFXANK
C	AR	Defective expression of MHC class II	RFX5
D	AR	Defective expression of MHC class II	RFXAP
CD3 deficiency			
γ	AR	Mild T-cell ID	CD3γ
ε	AR	Mild T-cell ID	CD3ε
Activation deficiencies			
Low CD8/abnormal CD4	AR	Functional T-cell ID	ZAP-70
Defective Ca^{2+} flux	AR	Functional T-cell ID	?
CD25 deficiency	AR	Functional T-cell ID	CD25
Other	AR	Defective tyrosine phosphorylation	?
Lymphokine deficiencies			
IL-2	?	Functional T-cell ID	?
Multicytokine	?	Functional T-cell ID	
		Abnormal NF-AT activity	?
PNP deficiency	AR	Progressive T-cell ID	PNP
		Neurological manifestations	
ADA deficiency (late onset)	AR	Progressive T-cell ID	ADA
Omenn syndrome	AR	Activated T cells in skin and gut Oligoclonality	RAG1, RAG2
XL T-cell deficiency	XL	Progressive T-cell ID	Mild type IL2Rγ of XL SCID

AR=Autosomal recessive; XL= X-linked recessive

While discussing this group of diseases, the earlier discussion on the activation of T cells and how they mediate their effector functions in chapter 7 should be kept in mind.

class I molecule and CD8 lymphopenia. The molecular defect is due to mutations in the HLA class I-associated peptide transporters TAP1 or TAP2 genes. These are located in the MHC region of chromosome 6 and transport peptides from the cytosol to the endoplasmic reticulum, where peptide loading of the class I molecule occurs. In the absence of TAP complex, MHC class I molecules are peptide free and unstable.

CD3 deficiency and CD8 deficiency

These are recently reported, rare conditions in which there are defects in the surface expression of CD3/CD8. This results in low anti-CD3 and antigen induced activation of T cells. However, B cell functions are not adversely affected and the response to vaccination is usually normal. Mutations have been found in the CD3 γ and ε chain encoding genes.

T cell activation deficiencies

Defects in the early steps of T cell activation have been found in some cases resulting in the typical presentation with opportunistic infections, recurrent viral or bacterial infections, diarrhoea, hepatitis and dermatitis and development of lymphomas. Defects may be due to low calcium flux, CD25 deficiency or defect in for Zap 70, a tyrosine kinase involved in TCR signalling.

Lymphokine deficiencies

These are rare conditions in which the lymphocyte numbers are normal but they have impaired ability to produce IL-2 or its receptor CD25 resulting in defective T cell proliferation. The exact nature of the defect has not been characterized.

T-cell deficiencies with associated features

The DiGeorgeSyndrome

Faulty development of the third and fourth brachial arches during the early foetal growth, leads to thymic, parathyroid and cardiac defects. Developmental changes can affect the face and parts of the brain. The abnormalities differ from child to child depending on which part is being affected and can range from mild to severe. Some children with DiGeorge anomaly have a distinctive look, with an under developed chin, eyes that slant downwards, and misshapen ear lobes. Some children have underdeveloped parathyroid gland resulting in unbalanced calcium levels and convulsions. Many children have a very small or absent thymus which results in defective cell mediated immunity. Autosomal dominant inheritance has been reported and cytogenetic abnormalities are found in most cases with deletions in chromosome 22q11 in approximately 90% cases.

The thymic defect results in T cell maturation defect and low or absent number of T cells in peripheral blood. T cells also show a poor proliferative response to mitogenic or allogenic stimuli while the antibody levels are normal. These patients are susceptible to mycobacterial, fungal and viral infections. Surprisingly, these children show a gradual improvement in their T cell function with age and it may even become normal by 5 years. It is not clear whether the T cell maturation occurs in the rudimentary thymus or extra thymic tissues take up these functions.

Short-limb dwarfism or Cartilage hair hypoplasia

This is an autosomal recessive condition characterized by short limbs, short stature and thin sparse hair. The skin forms extra folds around the neck, hands and feet, and the joints are loose. There is variably decreased number of T cells with the increased frequency of infections as in other T cell deficiencies.

Purine nucleoside phosphorylase (PNP) deficiency

PNP deficiency causes a progressive cellular immunodeficiency associated with defective antibody production in one third of cases. The disease may develop only in early adult life and is associated with autoimmune diseases in one third of cases. More than half the cases also have neurological manifestations consisting of mental retardation and developmental delays. This is an autosomal recessive condition and diagnosis is based on low PNP activity in red cells or lymphocytes. The T cell functional defect is believed to be the result of dGTP accumulation in the cell.

T cell deficiencies with cytogenetic abnormalities

Ataxia-telagectasia: This is an autosomal recessive disease characterized by the triad of features, recurrent sino-pulmonary infections, progressive involvement of the cerebella cortex manifested by an ataxic gait and fine vascular malformations, which are termed telangecatsia. The latter two features give this condition its name. Serum alpha-fetoprotein levels are usually elevated. Some form of immunodeficiency is found in over two third of cases although the defects are not consistent in all patients. These may include decrease in the serum level of IgG_2, IgA and IgE, reduced antibody response to protein and polysaccharide antigens and decreased number and function of T cells. The cells of these patients have hypersensitivity to ionizing radiation and lymphocytes show frequent chromosomal breaks, inversion, and translocations in the region of the T cell receptor and immunoglobulin genes. These patients as well as their first-degree relatives have a significantly higher risk for developing malignancies.

The cytogenetic abnormalities, predominantly in the T cells, are a hallmark of this condition. These include chromosomal rearrangements involving 7p14, 7q35, 14q12 and 14q32 which are the genetic locations of TCRγ, TCRβ, TCRα/δ and immunoglobulin heavy chain loci. The AT locus itself is at chromosome 11q23 and the gene codes for a protein that is homologous to the PI-3 kinase family. ATM protein is involved in controlling genomic stability and majority of patients have mutations in the AT gene.

A variant of AT is the Nijmegen breakage syndrome (NBS), which has similar chromosomal rearrangements in T cells, defective inhibition of DNA synthesis following irradiation and an immunodeficiency; however, no clinical signs of ataxia and telangiectasia are present. The NBS gene is also involved in maintaining genomic integrity and is distinct from AT gene.

Immunodeficiency, centromeric instability and facial anomalies (ICF syndrome)

This is a rare syndrome in which the immunodeficiency is of both T and B cells and the severity is variable. Cytogenetic studies show elongation of secondary constriction, which can be associated with a multibranched configuration predominantly in chromosome 1, 9, 16 and 2. Mutations in both alleles of the DNA methyl tranferase DNMT3B-encoding gene have been found in a number of affected families. The relationship of the mutations to immunodeficiency is not clear.

Complex syndromes

Wiskott-Aldrich syndrome: This is an X-linked disease, which presents in infancy or early childhood and is characterized by an immunodeficiency associated with thrombocytopenia, abnormal platelets and eczema. The platelets are typically small and show impaired aggregation. The defect is in the stability of the surface sialoglycoproteins CD43 and gp1b on leukocyte and platelet membranes. The cytoskeleton in T cells and platelets is abnormal and T cells show an abnormal proliferative response to anti-CD3. Serum Ig may be initially normal but decreases progressively. Antibody production, especially to polysaccharide antigens, is deficient. Along with this a progressive lymphopenia also develops which is predominantly of the T cells. This results in defective CMI status. Death occurs in early life often due to lymphoreticular malignancies. The risk of cancer has been calculated to be 2% per year. The gene defect has been localized to Xp11.22 and is non-randomly lionized during differentiation of blood cells making carrier detection possible. The WAS gene codes for a protein of unknown function.

Inherited complement deficiencies

Deficiencies of classical and alternative pathway include C1q, C1r, C4, C2, C3, properdin and Factor D deficiencies. Of these, C2 deficiency is the most commonly identified human complement deficiency, 1% of the population having heterozygosity for the deficiency and 0.01% complete absence. Complete deficiencies of the early classical pathway components (C1, C4, C2) are associated with a high risk of systemic lupus erythematosus (SLE) or other autoimmune disease (approximately 30% for C2 to >90% for C1); the risk for recurrent infections is less. C1 deficiency results in decreased levels of haemolytic complement activity and may be due to defects in C1q on chromosome 1 or the closely linked C1r and C1s genes on chromosome 12. C3 deficiency is associated with frequent serious pyogenic infections that may be fatal.

Individual who lack the terminal complement components, including C5, C6, C7, C8 and C9 are not able to generate membrane attack complex and would therefore be expected to be incapable of killing foreign organisms. However, it has been seen that patient with infections from N. *meningititis* and N *gonorrhoae* are deficient in C5, C6, C7 and C8. Interestingly, individuals with C9 deficiencies do not show the same susceptibility to infections suggesting that C9 is not necessary for *Neissseria* bacteriolysis.

Complement regulatory proteins are associated with abnormal complement activation and variety of related clinical abnormalities (**Table 9.6**).

Genetic deficiency of mannose binding lectin (MBP) in the lectin pathway and Factor D and properdin in the alternate pathway have also been reported. These individuals also have an increased risk of infections with encapsulated organisms.

Deficiencies of the complement receptors (CR) 3 and 4 are caused by mutations in the shared beta chain of these heterodimeric molecules. Patients typically present with recurrent pyogenic infections.

The genetic deficiency of the complement regulatory protein C1-INH is responsible for hereditary angioedema (HAE). The disease is characterized by non-infiammatory submucosal and subcutaneous edema. Edema most frequently occurs in the skin, respiratory tract and gastrointestinal tract. The disease may present for the first time only in late adolescence or even in adulthood. In most cases no precipitating factor may be found. The laryngeal edema may be severe enough to be life threatening. The inheritance is autosomal dominant and affected persons have low C2 and C4 levels between and during attacks of edema due to uncontrolled cleavage of C1s.

Table 9.6 Hereditary complement deficiencies in Humans

Affected	Frequency	Manifestations
CIq	F	Associated with hypogammaglobalinemia and severe combined deficiency.
CIr	1	Kidney and skin lesion, glomerulonephritis, recurrent infections lupus like syndrome.
CIs	I	Found in SLE
C2	Fb	Often no sign, increased in SLE and other autoimmune diseases.
C3	I	Frequent infections
C4	I	Lupus like syndrome
C5	I	Recurrent infections, lupus-like syndrome.
C6	I	No disease.
C7	I	No disease.
C8	I	Found in a patient with in gonococcal infection
CINA	F	Hereditary angioneuritic edema.
C3b INA	I	Frequent infections.

F = relatively frequent b = Most frequently observed deficiency
I = relatively infrequent

Inherited abnormalities of complement regulatory proteins

A number of well recognized clinical syndromes have been reported due to inherited abnormalities of complement regulatory proteins (**Table 9.7**). Hereditary Angioneurotic Edema is an autosomal

Table 9.7 Abnormalities / Deficiencies of Complement Regulatory Proteins

Protein	Resulting Complement Activation Abnormalities	Associated Disease Pathology	Disease Syndrome
C1 inhibitor	Deregulated classical pathway activation and consumption of C3	Acute intermittent attacks of skin and mucosal edema. Systemic Lupus erythematoses: B cell lymphoproloferative disease.	Hereditary angioneurotic edema
Factor I	Deregulated classical pathway activation, consumption of C3	Pyogenic infections, immune complex disease	
Factor H	Deregulated classical pathway activation, consumption of C3	Pyogenic infections, glomerulonephritis.	
Decay accetrating factor (DFA)	Deregulated C3 convertase activity.	Complement mediated intravascular hemolysis	Paroxysmal noctornal hemoglobinuria
Homologous restriction factor (HRF) CD59 (MIRL)	Increased Susceptibility of erythrocytes to MAC-mediated lysis.	Complement mediated interavascular hemolysis	Paroxysmal nocturnal hemoglobinunia

dominant disorder, characterized by edema of skin and mucosa. The skin of face and extremities and the laryngeal and intestinal mucosa are most frequently involved. The symptoms include severe abdominal pain, nausea and vomiting, diarrhoea, and potentially life threatening airway obstruction. Attacks have been associated with antecedent emotional stress and trauma. The physiologic basis of the association is not understood. The deficiency in C1INH patients may be either due to an absolute reduction in C1INH protein due to inheritance of the non-functional gene or by synthesis of a normal amount of a dysfunctional protein due to mutations in the coding sequences of the gene.

Deficiencies of the soluble alternative pathway regulators (factor I and factor H) are rare and the clinical consequences appear to include increased infections with pyogenic bacteria. These deficiencies have provided insights into the role of factor I mediated cleavage of C3b. Plasma C3 is completely consumed in these patients as a result of unregulated formation of fluid-phase C3 convertase. This is similar to the effects of an autoantibody called C3 nephritic factor (C3 NeF), which is specific for alternative pathway C3 convertase (C3bBb). This antibody stabilizes C3bBb and protects the complex from factor H-mediated dissociation. This results in unregulated consumption of C3. Patients with this antibody often have glomerulonephritis, possibly caused by inadequate clearing of circulating immune complexes.

Deficiencies in integral membrane protein regulators of complement activation include the absence of DAF, HRF and CD59. All of these proteins are normally bound to the plasma membrane of erythrocytes and other cells types by phosphatidylinositol linkages. Paroxysmal nocturnal haemoglobinuria (PNH) is a disease in which cells lack the ability to express phosphatidylinositol-linked membrane proteins including DFA, HRF and CD59. PNH is characterized by recurrent bouts of intravascular haemolysis, at least partially attributable to complement activation on the surface of erythrocytes.

Deficiencies in complement receptors include the absence of CR3 and CR4, both resulting from rare mutations in the chain (CD18) gene common to the CD11 and CD18 family of integrin molecules. The congenital disease caused by this gene defect is called **Leukocyte Adhesion Deficiency**. This disorder is characterized by recurrent infections, probably due to a combination of impaired iC3b-dependent phagocytosis of bacteria and inadequate adherence of neutrophils to endothelium at tissue sites of infection.

Miscellaneous defects

Chronic mucocutaneous candidiasis

Candida infections can persist in the skin and mucosae in patients with selective T-cell deficiency to *Candida* antigen. This susceptibility may be inherited as an autosomal recessive or dominant trait.

X-linked proliferative syndrome (XLP)

EBV infection provokes uncontrolled polyclonal, oligoclonal or monoclonal B cell proliferation in patients with T-cell deficiencies such as SCID. The gene has been mapped to Xq24-q27.

Phagocyte deficiencies

Disorders of neutrophil production

Chronic neutropenia is defined as absolute neutrophil count (ANC) of less than 500/μl lasting for more than 6 months. Genetic chronic neutropenia occurs in a number of syndromes and is a cause for recurrent pyogenic infections. These should be considered when the many common causes for infections have been excluded. Severe congenital neutropenia or Koštmann syndrome, Scwachman-Diamond syndrome, primary autoimmune neutropenia and cyclic neutropenia are members of this group of conditions.

Defect of Leukocyte adhesion

The migration of leukocytes to the site of infection is a crucial event in the initiation of inflammation. This is a complex event, involving the integrin molecules on the leucocytes, adhesion molecules on the endothelial surface and glycosaminoglycans or selectins. Leukocyte adhesion deficiency 1 (LAD 1) deficiency results from the mutations in the gene encoding the common β chain CD18 (common to the iC3b receptors CR3 and CR4 on phagocytes and the LFA -1 adhesion molecule of T cells), located on chromosome 21q22.3. This group of patients have severe life-threatening bacterial infections affecting skin, gingival, gut and lungs.

Chronic Granulomatous disease (CGD)

It is the name given to the group of inherited immunodeficiency diseases caused by faulty phagocytes. It is due to defect in the oxidative reaction in the phagocytic cells by which NADPH is converted to $NADP^+$ $2O^-{}_2$ + H_+ This failure to produce super oxide radicals, oxygen singlets and hydrogen peroxide results in the failure to kill intracellular organisms. The abnormality of the reaction is due to either a defective NADPH protein or defective 22 kd or 91 kd chain of the cytochrome b_{558}. The genes for these proteins have been localised and therefore CGD is a syndrome complex in which there might be a variety of defects. The affected children have inability to eliminate intracellular bacteria. There are four different types of CGD, each caused by different gene defect. Each of these genes encodes one of the four proteins that act together to allow phagocytes to kill bacteria. One gene is on chromosome X while other genes are recessive and are on autosomes. Infections in these children are often serious and even a harmless fungus like *Aspergillus* may cause skin infections and rashes, liver abscesses, fever and persistent cough. The diagnosis is established by the inability of neutrophils to reduce the dye nitroblue tetrazolium

Neutrophil G6-PD deficiency

This condition manifests just like CGD but is due to mutations in the G6-PD gene located at Xq28.

Chediak-Higashi syndrome

This is a rare autosomal, recessive disorder characterized by partial oculocutaneous albinism (lack of pigment in skin and eye), increased susceptibility to infection, deficient natural killer cell activity and abnormal giant primary granules in neutrophils.

Hyper-IgE recurrent infection (HIES or Job) syndrome

This is a multisystem autosomal dominant disorder characterized by recurrent infections of the lower respiratory tract and skin, chronic eczema, eosinophilia and very high IgE levels.

Defects of the interferon-γ pathway and monocyte function

The monocyte/macrophage system plays an important role in the host defence against mycobacteria and other intracellular pathogens. A group of genetic abnormalities in the IFN-γ receptors have been described which lead to infections with atypical mycobacteria. These include complete deficiencies of IFN-γRI and IFN-γRII as well as partial deficiency of IFN-γRI and dominant-negative IFN-γRI deficiency. Often these defects are noticed when BCG vaccination to neonates cause widespread dissemination.

Deficiency of IL-12p40 and IL-12 receptor β1 have also been reported. These are rare and also lead to increased susceptibility to mycobacterial infections.

Signs and Symptoms of primary immunodeficiencies (PI)

The most common clinical presentation of persons with PI disease is an increased susceptibility to infection. These persons suffer from repeated episodes of infections, which are, severe, long lasting, or difficult to treat. For example, healthy young people frequently suffer from cold, cough, and earache and many normal infants and young children may have one to three ear infections per year. On the other hand, children with PI, have repeated episodes of infection and may have two or three different infections at the same time. Weakened by repeated infections, these children fail to gain weight or fall behind in growth and development. Infections in this setting respond poorly to antibiotics and tend to become persistent or chronic. Children may suffer from chronic sinusitis (infection and inflammation of the sinuses, air passages in bones of the cheeks, forehead, and jaw) or chronic bronchitis (infection and inflammation of the airways leading to the lungs).

Serious infections, especially bacterial infections, result in repeated hospitalization. These may be pneumonia (an infection of the smallest airways and airsacs in the lungs, which prevents oxygen from reaching the blood), meningitis (infection of the membranes that surround the brain and spinal cord, causing fever and severe headache, and can lead to seizures, coma, and even death), osteomyelitis (infection of bone) and cellulitis (a serious infection of connective tissues just beneath the skin). Serious infection may result in septicemia (infection that flourishes in the bloodstream and spreads rapidly through the body) which is life-threatening.

Infections with organisms that do not usually invade individuals with a healthy immune system are known as "opportunistic" infections. These organisms take advantage of the opportunity afforded by a weakened immune system. For example, *Pneumocystis carinii* is a microscopic parasite that infects many healthy people without making them sick. But when the immune system is compromised, *Pneumocystis* can produce a severe form of pneumonia. *Toxoplasma,* another common parasite that usually produces no disease, causes life-threatening infection of the brain in persons with a weakened immune system. Besides infections, as has been seen in the above discussion of individual syndromes, many immunodeficiency diseases are associated with autoimmune disorders. The autoimmune manifestations can involve a single organ (pancreatic islet cells causing diabetes) or multiple organ systems. Finally, an immunodeficiency disease may be a part of a complex syndrome, with a tell tale combination of signs and symptoms. For example, children with DiGeorge anomaly not only have an underdeveloped thymus gland (and a corresponding lack of T cells), they typically have congenital heart disease, malfunctioning, or underdeveloped parathyroid glands, and characteristic facial features. Young boys with Wiskott-Aldrich Syndrome, in addition to being prone to infections, develop bleeding problems and a skin rash.

How to diagnose primary immunodeficiency

Sometimes the signs and symptoms of PI are so severe, or so characteristic, that the diagnosis is obvious. In most cases, it is not clear if a long string of illnesses are just "ordinary" infections, or if they are the result of an immunodeficiency.

1. Many conditions can produce an immunodeficiency, at least temporarily, and most children who seem to have "too many" infections are not, in fact, suffering from an immunodeficiency. Experts estimate that half of the children who see a doctor for frequent infections are normal, another one third may have allergies, and 10 % have some other type of serious disorder. Only 10 % turn out to have a primary or secondary immunodeficiency and this should be suspected in any of the following conditions.

- Multiple episodes of ear infections within a year
- Two or more serious sinus infections within a year
- Two or more months on antibiotics with little effect.
- Two or more pneumonias within a year
- Failure of an infant to gain weight or grow normally.
- Recurrent deep abscesses in the skin or organs.
- Persistent thrush in mouth or on skin, after one year of age.
- Need for intravenous antibiotics to clear infections.
- Two or more deep-seated infections such as meningitis, osteomyelitis, cellulitis, or sepsis.

When a pattern of frequent infections suggests an

immunodeficiency, the doctor begins by exploring the patient's "history" and the family's history, and then conducts a physical examination.

- *The patient's history*. What infections have the patient had in the past, or have now? Have they been unusually frequent, or severe, or long-lasting? Have they failed to respond to standard treatments? When a child who is immunologically normal develops a string of infections, they are usually mild and short-lived, and between infections the child recovers completely.

What, besides a PI, might explain the high rate of infections? Normal immune responses can be suppressed by many factors, including malnutrition, injuries such as burns, and certain types of drugs (corticosteroids, for instance). Immune responses can also be suppressed by some diseases, such as leukemia, and some infections, such as infectious mononucleosis (mono), measles, chicken pox, and AIDS. In fact, almost every serious illness impairs the immune response.

- *Physical examination:* Is the child well-nourished and growing well? A severely immunodeficient child is likely to look sick and pale. Very often the child is underweight and lags behind in growth and development.

The child may be shy or quiet. An active, robust, healthy-looking child is less likely to have a serious immune deficiency.

The doctor will listen for changes in the lungs and look for rashes, sores, thrush in the mouth, an enlarged spleen or liver, and swollen joints. Some immunodeficient children may lack palpable tonsils or lymph nodes in the neck.

- *Family history*. Have any family members or relatives ever been diagnosed with PI or shown an unusual susceptibility to infections? Have there been any infant deaths from infections? Were only boys affected?

Evaluating immune response

To decide if an illness can be traced to an immunodeficiency, laboratory investigations are necessary. These tests probe the soundness of the various components of the immune system. Are all the right immune cells present, in adequate numbers, and are they working properly? Are there normal amounts and types of antibodies?

The screening tests are relatively simple and inexpensive routine tests. Two routine tests - complete blood count and quantitative immunoglobulins - will detect most, but not all, immunodeficiencies.

If antibodies are normal, or if the patient's infections seem to be caused by viruses or fungi, T cells should be checked. If T cell numbers are normal and function normally, phagocyte function should be evaluated.

The most common screening tests include:

- *Blood count.* A complete blood count (CBC) will indicate the numbers of red and white blood cells as well as platelets. A "differential count" gives the proportion of different types of white blood cells, including lymphocytes and neutrophils.
- *Quantitative immunoglobulins.* This standard laboratory test measures various immunoglobulin levels in the blood. In addition to total immunoglobulins, it shows levels of the different immunoglobulin types (IgG, IgM, and IgA).
- *Antibody responses.* Is the response to an antigenic challenge normal? The antibody titer in the blood to the usual childhood immunizations, i.e., tetanus, measles, pertussis, or diphtheria gives a good estimate of the functional status. In older individuals, a booster injection to tetanus toxoid followed by estimation of the anti-tetanus respone is a useful approach.
- *Complement.* The components of the complement system, C3 and C4 are quantitated.
- *Skin tests.* These tests, which are similar to TB skin tests, show how well T cells are functioning. Small amounts of several standard antigens (including mumps and *Candida*) are injected into the skin. A person with a healthy immune system usually develops local swelling within 24 to 48 hours. However, these tests are not as accurate in very young infants.

When screening tests indicate an immunodeficiency—or when they fail to explain a stubborn infection—additional tests will likely be needed. There are dozens of sophisticated tests that allow doctors to identify and count subsets of B cells and T cells, and to assess subtle abnormalities in antibodies, immune cells and immune tissues. Tests can also probe the characteristics of infectious agents.

Evaluating infections

If an infection proves resistant to standard treatments, it is important to find out exactly what organisms are involved. Samples of mucus, sputum, or stool, or sometimes a small sample of the infected tissue itself are "cultured" and the organism identified making it possible to select the most effective treatment.

The infection itself often provides a good clue to the nature of an immune defect. Common bacteria typically elicit antibodies, while viruses and fungi stimulate T cells. Thus, sinus infections and respiratory infections, which are most often due to bacteria, suggest an antibody deficiency. Infections caused by a variety of viruses and fungi, or by *Pneumocystis*, point to a T cell defect. Recurrent infections involving the skin or soft tissues can often be traced to problems with phagocytes. Blood-borne infections caused by encapsulated bacteria, including meningitis, may be linked to complement deficiencies.

An experienced physician also finds clues to the diagnosis in the combinations of features. For example, a young infant suffering from diarrhea, pneumonia, etc, and exhibiting "failure to thrive," may have SCID. A 4-year-old with swollen lymph glands, skin problems, pneumonia, and bone infections may have Chronic Granulomatous Disease (CGD). A 10-year old with sinus and respiratory infections, an enlarged spleen, and signs of autoimmune problems is going to have Common Variable Immunodeficiency.

Gene therapy

Gene therapy is a new and revolutionary approach for correcting genetic defect and was first used to treat two young girls with Severe Combined Immunodeficiency due to ADA deficiency. Gene therapy attempts to cure disease by inserting a healthy version of a missing or malfunctioning gene into a cell to restore normal function. If successful, the newly-inserted gene directs the cell to produce the missing protein.

In the pioneering 1990 experiment, T cells were removed from the children, treated to make them more active and a gene for ADA was introduced. These manipulated T cells carrying the new gene were then injected back. These children continue to be healthy and free of severe infections and attend school, living relatively normal lives. However, the level of ADA enzyme produced by the transfused cells was inadequate and they needed supplementation with injections of ADA itself.

Efforts to improve the effectiveness of the new gene in the recipient are slowly yielding results. The gene is put into stem cells, which are much longer lived than T cells. In three different cases, babies were diagnosed with ADA deficiency before they were born. Their own umbilical cord blood was collected, and stem cells taken from the cord blood had new genes inserted. There babies were then given a transfusion of his/her own genetically -engineered stem cells. These children did well initially but, they continued to require other treatments also at later stages. In a similar trial in Fránce, the recipients developed leukemias and died. This was a result of the insertion of the gene into locations that affect the cell cycle. It is still not possible to control the site of gene insertion. However, many of these issues are being gradually resolved and the scientific community remains optimistic that successful gene therapy will become a reality in the foreseeable future.

CHAPTER

10

Transplantation Immunology

Transplantation is the replacement of a non-functional organ or tissue, such as kidney or bone marrow, with organs or tissue from a healthy donor so as to restore the essential function. Transplantation immunology is the study of immunologic reactivity of a recipient to the transplanted organ or tissues.

As far back as in the late nineteenth century, a Viennese surgeon observed that a kidney from an animal could be removed and put back surgically into the same animal without loss of function; however, when it was put into another animal there was loss of function. This was an important landmark in transplantation immunology, which along with the studies on mouse skin graft rejection led to the discovery of the major histocompatibility complex.

In the 1940s, Peter Medawar was the first to suggest that loss of function following a kidney allograft could be due to immunological reasons. He performed skin grafts in patients of burns and found that grafts from the same patient were well accepted; skin grafted from a donor was rejected. He also observed that following rejection of one graft, a second graft from the same donor was rejected in an accelerated manner. Since then great progress has been made in understanding the immunological principal of transplantation. This has led to the successful development of therapeutic agents to deal with allograft rejection and has made the current scenario possible. Today, a large number of successful transplants of kidney, liver, bone marrow, heart, lung and pancreatic islet cells, pancreas, cornea are being done every year. (**Table 1**) Of these, bone marrow is different from other organs (which are together termed 'solid organ'), since it is itself an immuno-competent tissue. We will discuss kidney transplantation as the classical example of a solid organ graft, since it is the most widely performed allograft. The special features of bone marrow graft will be briefly discussed.

Table 10.1 The number of organ transplantations performed in the United States of types of grafts

	1998	1999	2000	2001	2002	2003
Heart	2400	2200	2200	2200	2200	2100
Lung	850	900	975	1050	1000	1050
Heart and lung	45	50	48	28	32	29
Kidney	12100	12250	13600	14100	14800	15000
Pancreas	240	350	425	470	550	500
Kidney and pancreas	880	940	900	890	900	890
Liver	4500	4600	5000	5100	5200	5500
Intestine	70	74	80	110	105	115

Types of grafts

The donor recipient relationship defines the type of graft. Grafts between genetically identical individuals like monozygotic twins are termed an **isograft.** If the graft is between the members of the same species it is known as the **allograft** (allo meaning 'not self') and between different species is termed **xenograft.** .

Graft Acceptance and Rejection

Following the surgical grafting of an organ, revascularization and healing occurs rapidly with restoration of the functions. However, immune reactivity to the foreign donor antigens, except in the case of identical twins, results is clinical graft rejection. The severity of this allo-response depends on a variety of factors, which will be discussed in this chapter. It remains the single largest impediment tothe success in the field of transplantation.

Types of rejection

Hyper acute Rejection

Hyper acute rejection occurs within minutes to hours of transplantation. Immediately following vascular anastomosis a renal graft looks pink and starts to make urine. However, in hyper acute rejection, it rapidly becomes mottled and ischemic and fails to make urine, leaving the clinician no option but to remove the grafted organ. Histological examination of the rejected kidney shows fibrin and platelet thrombi, fibrinoid necrosis of the vessel walls with absence of any mononuclear cell infiltrate. Mechanisms of hyperacute rejection are well understood and occur because of the presence of anti-donor antibodies prior to the grafting. These bind to graft antigens on the vascular endothelium of the donor kidney. The antibodies may be directed against HLA antigens (usually HLA class I), the AB blood group antigens or the poorly defined endothelial antigens. This is followed by complement activation and its consequences.

Acute Rejection

Episodes of acute rejection commence 5 to 7 days after transplantation and occur with decreasing frequency after 3 months. It is often associated with withdrawal of immunosuppression. The major regulator of acute rejections is the T cell and in an experimental allograft model, elimination of CD4+ T cells prevents acute rejection. Consistent with these observations, most episodes of acute rejection after renal transplantation can be reversed with immunosuppressive agents directed at T lymphocytes, including high dose corticosteroids and the OKT3 monoclonal antibody, a monoclonal antibody directed against the T cells. Acute rejection is not homogenous and includes cell-mediated cytotoxicity by CD8+ cytotoxic T cells, DTH reactions induced by CD4+ T cells, and in some cases antibody mediated damage. The role of macrophages and natural killer cells is less well established.

Accelerated Rejection

Aggressive episodes of rejection occurring within 5 to 6 days of transplantation are termed accelerated rejection. It is believed to be due to prior sensitisation of the patient to antigens expressed by the graft. The kinetics of accelerated rejection is consistent with a memory or secondary immune response. Prior exposure to the donor graft antigens may be through blood transfusions, pregnancy or previously rejected grafts. The pathology is characterized by fibrinoid necrosis of the small vessels, which is consistent with an antibody-mediated process.

Chronic Rejection

This may occur from months to years following an allograft. The pathogenesis is multifactorial. The half-life of HLA identical live donor renal allografts is 25 years as compared to 8 years with cadaver donors. The immune mechanisms involved are poorly understood; however, it has been suggested that early phase of chronic rejection is immune mediated. The renal histology is characterized by increasing fibrosis. In addition to continued immune destruction non-immune factors including decreased nephron mass and hyper filtration also contribute to progress during the late phase.

Mechanism of graft rejection

The immune response following an allograft is primarily against major histocompatibility complex (MHC) molecules of the donor, which differ from that of the recipients. Approximately 8-10% of the normal adult T cell repertoire is capable of recognizing and responding to the foreign MHC molecules. This response is not to the benefit of the host. T cells recognize major histocompatibility complex antigens in transplantation by two-pathway i.e. direct pathway and indirect pathway (**Figure 10.1**). In the direct pathway, donor APCs bearing processed donor antigens are recognized by the recipient T cells. These APCs are transferred with the grafted organ and this form of recognition does not require processing and presentation by host (APCs) and is therefore, not MHC restricted. Because the frequency of T cells that are able to recognize alloantigen directly is very high,

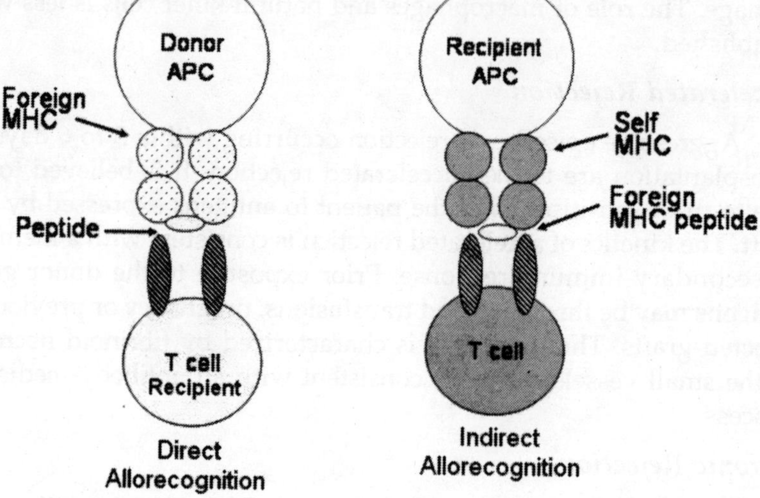

Figure 10.1 Mechanism of allorecognition. *Direct allorecognition involves recognition of foreign major histocompatibility complex (MHC) class I molecules with a variety of peptides present in the antigen presentation groove. Indirect allorecognition involves recognition of peptides derived from foreign MHC molecules and presented by self-MHC molecules and presented by self –MHC molecules and presented by self-MHC class II molecules.*

even in non-immunized responders, it is believed that this process reflects T cell recognition of allogenic MHC/peptide complex via molecular mimicry with other antigenic structures. Evidences in support of the direct recognition pathway in allograft rejection are based on the observations that depletion of donor APCs can some time prolong the allograft survival and that donor MHC is more important than minor antigens in causing graft rejection. Although the majority of T cells infiltrating the graft during early acute rejection exhibit direct recognition ability, it is unlikely that these cells can mediate late or chronic rejection because their stimulation requires the presence in the graft of passenger APC of donor. The absence of co-stimulatory molecules on the surface of graft endothelial and parenchymal cells renders such putative targets more likely to induce anergy rather than to stimulate recipient's T lymphocytes.

In contrast to the direct recognition pathway, reactivity of T cells to allogenic MHC peptides and proteins in the context of self-MHC is termed an indirect response. In this case peptides generated by proteolysis of allogenic MHC molecules bind to MHC- class II antigens of host APC and trigger T cell alloimmune responses. This form of allo-reactivity is restricted by host HLA-DR antigens and is carried out by an oligoclonal population of T cells, which are capable of recognising the dominant epitope of the allogenic MHC molecule. Because the stimulatory peptide can be generated continuously from soluble MHC alloantigens released from the graft and processed by host APC, it is generally believed that the direct pathway is responsible both for initiation and perpetuation of allograft rejection.

The exact mechanism by which T cells destroy the graft is not yet clear. Both CD4 and CD8 subclass of effector cells probably destroy graft cells by classical cytotoxic T cell mechanisms. Another important consequence of T cell activation is the release of other lymphokines, especially interferon (IFN- γ). IFN-γ induces increased expression of HLA-A, -B and DR on graft tissues thus potentially making the graft more vulnerable to effector mechanisms. IFN-γ also activates monocytes to mediate a destructive delayed hypersensitivity response against the graft.

In addition to IL-2 and IFN-γ released from activated T cells, IL-4 and IL-5 play a role in directing B cell for the production of antibodies. Antibody-mediated damage may then take place directly through complement activation or recruitment of antibody dependent cell mediated cytotoxic (ADCC) effector cells. Most of the cells that arrive in the graft in the early post-transplantation stage are lymphocytes, which migrate out of the capillary beds. After 7 days a heterogeneous collection of cell types are seen in the grafted tissue. Lymphocytes are the predominant cell type in this inflitrate, followed by monocytes/macrophages and few polymorphonuclear neutrophils. Thus both cellular and antibody mediated mechanisms are important in mediating graft rejection/acceptance.

T Cell mediated Rejection

It is well established that thymus derived T cells play an essential role in acute allograft rejection. Animals that are naturally or experimentally deprived of T cells (eg. nude mice, SCID mice, thymectomized mice) are unable to reject allografts. Following passive transfer of T cells into athymic mice, vigorous graft rejection takes place. In clinical transplantation, the successful treatment of rejection episodes with anti-T cell antibodies such as monoclonal anti-CD3 antibody (OKT3), antithymocyte globulin and antilymphocyte globulin has provided indirect support for the importance of T cells in graft rejection. More than 50% of rhesus monkeys treated with a CD3 immunotoxin, which lyses all CD3+ T cells, just before transplantation had long-term graft acceptance.

Both CD8+ and CD4+ T cells are activated by recognition of alloantigens of the grafts; the CD8+ T cells recognize foreign MHC class I molecules, which are expressed by all the cells in the graft (**Figure 10.2**). The differentiation of cytotoxic T lymphocytes (CTLs) is largely dependent on CD4+ T helper cells being stimulated by the indirect pathway of activation. The CD4 subset and its lymphokine products are the principle mediators of rejection. The most important APCs stimulating an antigraft response are the dendritic cells residing in the interstitium of the graft. Dendritic cells are regarded as critical instigators and regulators of immune reactivity, and play a key role in both direct and indirect pathways of allorecognition. Molecular signalling between dendritic cells and T helper cells directs the differentiation of naive (Tho) cells into either Th1 or Th2 cells.

Antibody Mediated Rejection

The role of antibody in hyper-acute rejection in renal transplant has been clearly established. A direct correlation is seen between positive pre-transplant cross-match which detects anti-MHC class I antibodies and the development of hyper acute rejection. Anti-graft antibodies can be eluted from donor kidneys after hyper acute

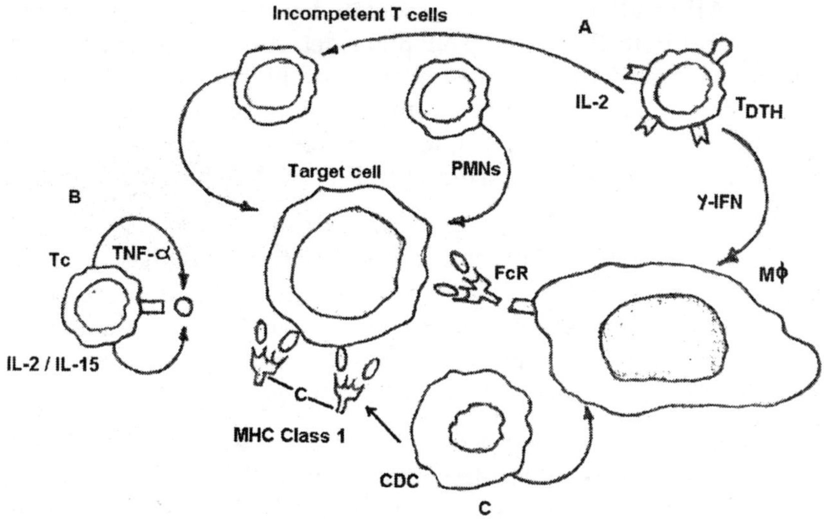

Figure 10.2 Mechanism of immune mediated injury of organ graft during rejection *(A) Cytotoxicity produced by delayed type hypersensitivity T cells (TOTH) recount granulocytes (PMN) and macrophages (Mϕ) to attack graft. (B) Specific T cells*

rejection. The passive transfer of antigraft antibodies in experimental models can provoke hyper acute rejection. It is likely that antibodies also play a role in other types of rejection; however, their mechanisms remain incompletely understood and also controversial especially in chronic rejection. Although the cellular infiltrate in most cases of chronic rejection is scanty, direct evidence for antibody-mediated damage is inconclusive. The antibodies causing hyper acute rejection may be preformed or they may develop later although patient is on immunosuppressive drugs, which could modulate the rate of production. Antibodies can bind to the graft, making the detection of soluble antigraft antibody difficult.

Role of cytokines and chemokines in allograft rejection

Cytokines are soluble mediators secreted by one cell that acts on another cell or organ. On recognition of MHC-peptide complex by the T cell receptor, the T helper (Th) cells differentiate into either Th1 cells that produce high levels of interferon (IFN -γ), IL-2 and TNF - α or Th2 cells that secrete high levels of IL–4, IL-5, IL-10 and IL-13. While IFN-γ and TNF-β promote cell mediated cytotoxicity and delayed type hypersensitivity reactions, Th2 cells promote humoral response. The Th1 and Th2 responses counter regulate one another. Allograft rejection is associated with activation of Th1 response and both IFN-γ and TNF-β are detected in rejected tissues. However, this is more often seen in skin and kidney grafts than liver or lung grafts. A Th1 to Th2 immune deviation can induce islet allograft tolerance across multiple minor histocompatibility antigen barriers. Th2 cytokines and mixed Th1/TH2 profiles have been shown in other types of grafts.

Genetic polymorphisms of pro-inflammatory cytokines and their regulatory elements have been identified. Some of these have functional implications and influence the quantitative and qualitative nature of the immune response. Thus, the genotype of the recipient may influence the severity of the anti-graft responses.

Chemokines also play a role in cellular rejection. These are a family of 8 - 11 kDa molecules involved in cellular traffic through tissues. More than 40 chemokines and over 18 chemokine receptors have been identified. There are distinct groups of chemokines; CC chemokines with cysteins adjacent to another, CXC chemokines with cysteins separated by one amino acid residue and CX3C in which cysteins are separated by more than 3 amino acid residues. In renal allografts, the expression of the chemokines MCP1, MIP - Iβ and IL-8 are increased in the early stages of rejection. This is in keeping with their role as chemo-attractants for inflammatory cells, especially polymorphonuclear cells. The detection of these in the urine could be a good marker for diagnosis of early rejection of renal allografts.

Response to bone marrow transplantation (BMT)

Most of the discussion regarding alloresponse has been in context of solid organ transplants. The transfusion of allogenic bone marrow as a source of stem cells poses a unique problem since the grafted tissue is itself immunologically competent. If the BMT is performed for congenital immunodeficiencies, the host is already immuno-compromised and hence accepts the graft very well. In many other situations such as leukemias, treatment of the disease results in total elimination of all the host immune cells. Transfusion of the donor stem cells, results in the donor cells mounting an immune response against the host organs. This is the reverse of allograft rejection and is termed graft-versus-host reaction or GVHD. This is a cell-mediated immune response that is mediated by CD4+ and CD8+cells. The clinical consequence of this is very severe with the patient having severe gastrointestinal and skin symptoms and can be life threatening.

Role of HLA matching in transplantation

Graft immunogenicity is a crucial factor for allograft rejection and HLA matched renal grafts outperform less as compared to well-matched allografts. Patient survival is also prolonged in recipients receiving better HLA-matched grafts. The donor and recipient should be matched for both the broad specificities that are detected by serological method and for the subtypes that are detected by molecular methods. The HLA system comprises multiple class I and especially class II genes; however, most matching strategies consider only HLA-A, HLA-B and HLA-DR loci. Although it make sense to find a perfect match for each transplant patient, if one considers the extensive polymorphism in these loci, it is impossible to find such donors. Hence, the best-matched donor is selected from among those available, keeping various clinical considerations in mind (**Table 10.2**).

Table 10.2 Graft survival on the basis of HLA matching.

Number of HLA mismatches	5 year renal allograft survival (%)
0	68
1	61
2	61
3	58
4	58
5	57
6	56

If it was possible to delineate the recipient's immune responder status, perhaps by genotyping for cytokines and other immuno-regulatory molecules, it may be possible to design strategies that combine tissue matching with the immune responder status and achieve better results. For example, well-matched grafts could be given to the high responders (to alloantigens) and the poorly-matched grafts to low responders.

Role of anti HLA antibodies in rejection

As has been discussed earlier, the presence of pre-existing anti-donor antibodies is associated with hyper acute rejection in solid organ transplants. This is of particular concern in renal grafts, since patients of chronic renal failure are often anemic and receive multiple blood transfusions. This sensitises them to various HLA antigens resulting in the presence of anti-HLA antibodies. Hence prior to any grafting, a cross-match is performed using the recipient's serum and the donor cells by the standard NIH-2 step microcytotoxicity assay, which has been described in chapter 3.

Some patients have antibodies directed against a wide range of HLA antigens, and these are detected when recipient's serum is cytotoxic to cells other than donor cells. These are termed panel reactive antibodies (PRA). Patients who are PRA positive are at increased risk of early graft loss.

Role of minor histocompatibility antigens (mi-HAgs) and tissue specific antigens

Minor histocompatibility antigens may play an important role in the graft rejection and are defined as cell surface antigens other than the MHC antigens. Linkage studies in congenic strains of mice have shown that mi-HAgs loci are scattered throughout the genome and there may be several hundred such antigens. Most of these loci are located on autosomal chromosomes; however, few have also been mapped to mitochondrial DNA and Y-chromosome. Because of this large number of genes, it is not possible to match these loci. It is also due to this reason that completely reaction free graft acceptance only occurs between identical twins.

MiHAs are not universally present on all the cells and they don't interact functionally with MHC antigens. These antigens are less potent and immunogenic than the HLA antigens and do not initiate the immune response independently The immune response to mi- HAgs is MHC restricted and T cell mediated, predominantly by cytotoxic T lymphocytes. Since most of these antigens are intracellular proteins, no antibody response is seen. They may be involved in the comparatively slower, chronic form of rejection. .

Tissue specific antigens are antigen systems that are expressed only on one type of organ, tissue or cell and are independent from antigens such as HLA antigens, which have a wide distribution throughout the body. It has been observed that different organs grafted from the same donor behave differently. For instance while porcine skin and kidney grafts which may get rejected, liver allograft survives in unrelated pigs. Several cases of multiple organ transplants have been reported in which one organ is rejected while the other continues to function. These observations suggest that tissue specific antigen systems may play a role in graft acceptance and rejection.

If all relevant transplant antigens were ubiquitous, HLA compatibility between recipient and donor may prevent sensitization to tissue specific antigens. However, HLA identity and negative MLC does not prevent immune response to the increasingly well-studied vascular endothelial cell (VEC)-specific antigens. The vascular endothelial cells of transplanted organs are at the interface between the graft and the recipient's blood containing immuno-competent cells. VEC antigens are expressed in abundance throughout the renal vasculature and in extra renal vasculature. The VEC synthesize clotting factors and inactivating hormones and can act as antigen presenting cells. Hence, VEC plays an important role in the rejection process and anti-VEC antibodies are the most commonly encountered antibody in-patients rejecting a renal allograft. It is detected in 96% of the patients who experience chronic rejection. Most patients with anti-VEC antibody do not have any anti HLA antibody to the donors. Antibody to VEC is rarely encountered in normal persons and in low frequency inpatients experiencing a benign clinical post transplantation course.

Other factors involved in transplant rejection

Heat shock proteins (HSP) have also been implicated in the pathogenesis of chronic rejection. HSPs are ubiquitously present and act as chaperones for protein traffic. They are also upregulated during the stress response. The members of the hsp 70 family have a common structure consisting of a C - terminal peptide-binding domain and a N-terminal ATPase domain which influences peptide binding. For instance, grp 78 can bind peptide of minimally 7 –8 residues and the peptide-binding region has four major pockets that can accommodate large hydrophobic residues. Through its peptide-binding properties, grp 78 might participate in the transport and processing of autologous peptides presented by APC through alternative pathways of antigen processing and presentation. Tissue injury leads to an accumulation of protein degradation products and elevated levels of heat shock proteins that function as chaperones in protein renaturation and in other cytoprotective processes. The degradation of denatured protein generates (auto) antigenic peptides that bind to heat shock proteins such as grp 78, grp 94 and PDI and such complexes translocate into cellular compartments involved with antigen processing and presentation. Thus, stress responses may activate hsp-dependent pathways of (auto) antigen-induced T- cell activation.

Current experimental evidence supports the concept that during cellular rejection, graft infiltrating cells induce a stress response within the allograft which increases the expression of heat-shock proteins and triggers the recruitment and activation of hsp-dependent lymphocytes. Number of stress proteins have been shown to have higher tissue levels during different phases of allograft rejection. However, how much is the contribution of these mechanisms in clinical transplant rejection is not clear.

When APCs are stressed by heat, chemical treatment or ultraviolet irradiation, they lose their ability to stimulate in mixed leukocyte cultures. It has been shown that animals injected with stressed allogenic cells often exhibit prolonged graft survivals and even transplant tolerance. Since stressed cells have increased hsp expression, it is possible that these effects might be related to hsp-mediated immunomodulation.

Prevention and Treatment of Allograft Rejection

Except in the case of transplantation between identical twins, all allografts result in variable degrees of immune activation. Allograft recipients are maintained on long-term immunosuppressive regimes to give the grafted organ the best chance to remain functional. Much progress has been made in recent years in the development of new immunosuppressive molecules. However, despite this progress, these agents are at best non-specific suppressors of immune reactivity. Since they often globally alter the immune system, there is an increased risk of infection and malignancies. There is also the danger of the adverse effects of these drugs on non-immune organs due to the high doses that are required to achieve adequate immunosuppression. In fact, in some instances the adverse response to the drug is more harmful to the patient than the graft rejection for which the drug is being given. The clinician may in such instances withdraw the drug and allow the graft to be rejected in order to save life.

Since graft rejection is a predominantly T lymphocyte mediated immune response to the grafted organ, the response can be theoretically blocked at any level in this process. (**Figure 10.3**) We will deal briefly with the various strategies that can be used to achieve immunosuppression and highlight the drugs and biological material that are being currently used in transplantation.

Interruption of allorecognition: The first level of intervention could be abrogating the initial events in the alloresponse such as leukocyte migration into the graft interstitium. Although blocking adhesion molecules with antibodies to LFA-1, ICAM-1 have had some success in experimental models, the results in human renal transplantation have been poor.

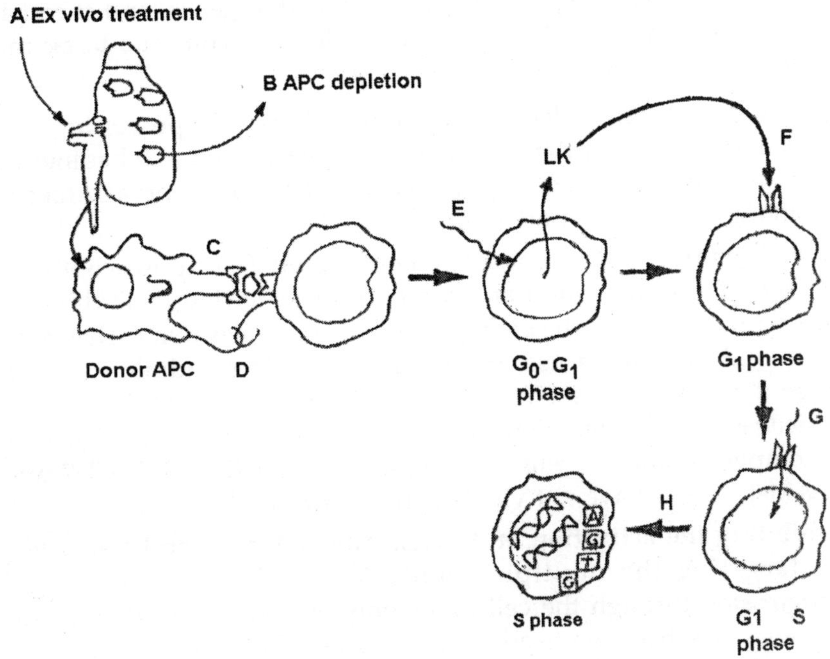

Figure 10.3 Possible sites at which immune intervention can be targeted. *(A) Ex-vivo treatment of graft to reduce ischemia perfusion injures (B) Depletion of antigen presenting cells to reduce immunogenicity (C) Dampening of donor APC activity (D) Disruption of co receptor interactions (E) Inhibition of antigen – driven events that result in up regulation of inflammatory cytokines (F) Blockade of cytokine receptors (G) Dampening cytokine receptor signal transduction (H) anti metabolites to block nucleoside synthesis*

Depletion of graft antigens: The elimination of passenger dendritic cells from the grafted organ would abrogate chances of direct allorecognition. This can be achieved either by treating the donor with corticosteroids and cyclophosphamide or treating the isolated graft with irradiation or pan-leukocyte antibodies. However, these approaches have not been successful in renal grafts.

Inhibition of antigen presentation: Steroids are used to dampen antigen presentation by APCs. Steroid molecules enter the cytosol and bind to steroid receptors, migrate to the nucleus and binds regulatory regions of DNA termed glucocorticoid response elements.

These regulate the transcription of many genes, explaining the effects of steroids on multiple cell types and functions. This accounts for both its efficacy and the diverse array of side-effects. Steroids also have multiple effects on the immune system. They block T cell activation cascade, expression of IL-1 and IL-6 genes, and through indirect mechanisms IL-2 production. Steroids continue to be the mainstay of all immunosuppressive regimens.

A new drug, 15-deoxyspergualin inhibits APC processing and presentation, thereby dampening the immune response. It has shown only limited clinical activity. However, this is an approach that is receiving considerable attention.

Blocking the T cell receptor and co-receptor: This approach would prevent T cell activation and a monoclonal murine IgG2 directed against the ε chain of the pentapeptide CD3 complex (OKT3) has been approved for human use. However, it need to be administered in large doses to achieve effective T cell silencing and at this dose the toxicity is considerable. Hence this drug is given to treat acute rejection. Attempts to block T cells with antibodies to CD4, CD2, CD7 and ICAM-1/LFA-1 have not yielded any useful results.

Inhibiting the activation of T cells: Although this can be achieved by targeting the multiple events involved in activation and progression through the cell cycle, only one successful therapeutic strategy has been devised. This involves the inhibition of Ca^{2+}-dependent cascade which generates the regulatory proteins that control DNA transcription of a wide range of mRNA. These include the proinflammatory cytokines IL-2, IL-13, IL-4 and IFN-γ, all of which play an important role in rejection. Cyclosporine and tacrolimus (FK 506) are two compounds of fungal origin that act by this mechanism. These compounds are hydrophobic in nature and easily penetrate plasma membranes and bind to the cytoplasmic cis-trans prolyl isomerases, cyclophilin and FK- biding protein respectively. These complexes inhibit the enzymatic actions of calcineurin to dephosphorylate and transport the nuclear factor of activated T cells (NF-AT) to the nucleus. This is the essential initiating event in the regulation of cytokine genes.

The introduction of these agents has dramatically altered the outcome of renal transplantation and brought down the frequency of acute rejection episodes from 85% to 50%. Tacrolimus has been shown in vitro to be 25 to 50 fold more potent than cyclosporine.

Interfering with cytokine receptors: Various types of monoclonal reagents have been developed to block the IL-2 receptor. However, they have been found to have limited effectiveness in patients.

Inhibiting the cytokine signal transduction: Sirolimus (Rapamycin) is another fungal metabolite, which is structurally related to FK-506 and binds to same proteins. However, its mechanism differs from that of FK-506, since it prevents phosphorylation of a p70 kinase in the CD-28 co-stimulatory and IL-2R signal transduction pathway. Rapamycin blocks T cell proliferation during late G1 and pre-S phase of the cell cycles (**Figure 10.4**). Thus rapamycin inhibits late signals in T cell activation while cyclosporine and FK-506 inhibit an early signal in T cell receptor signal transduction pathway and these compounds have synergistic effect when given together. Activation of T cells with monoclonal antibodies is inhibited by cyclosporine and FK-506 but not by rapamycin.

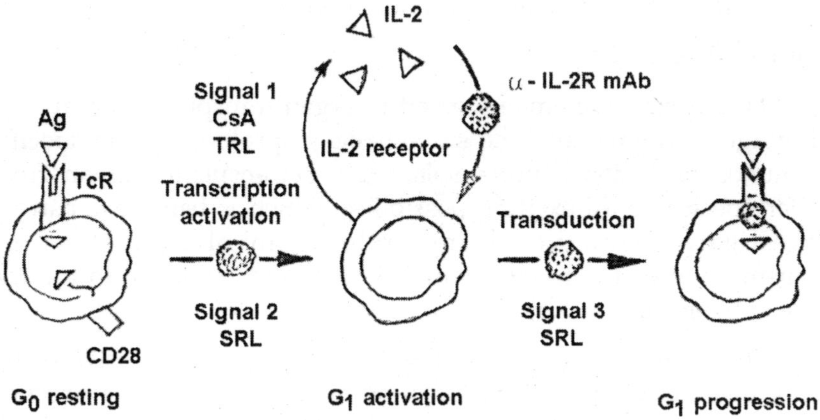

Figure 10.4 Complementary sites of action of immunosuppressive drugs during lymphocyte activation. *Cyclosporine (CSA) and tacrolimus (TRL) inhibit transcription of T- cell growth promoting genes like IL-2. Anti-IL-2mAb block binding of IL-2 to its receptor Sirolimus (SRL) blocks growth factor initiated signal transduction.*

Inhibiting nucleoside synthesis: The purine salvage pathway in lymphocytes is less active than in other rapidly dividing cells. Hence, these cells are more dependent on the de novo synthesis of DNA and RNA building blocks. Inhibition of de novo nucleoside synthesis achieves a milder immunosuppression than with CsA or Rapamycin and it also inhibits many other cell types. Compounds, which act by this mechanism, include azathiprine, mizorbione and mycophenolic acid.

Anti-lymphocyte globulin: More general immunosuppression is achieved by use of anti-lymphocyte globulin, which has an overall lymphocyte depleting property. This has been licensed for use for a long time and was widely used prior to the availability of the more specific agents like OKT3 and anti-IL2R antibodies.

Excellent theoretical rationale and successful use in animal models of transplantation do not assure a good product for use in the clinics. The toxicity profile becomes an overriding concern and treatment regimes are constantly evolving in order to achieve the best immunosuppression and resultant graft survival with least side-effects.

The effort is to provide intense suppression in the first 2 weeks post-transplant in order to minimise acute rejection. The patient has to be maintained permanently on low dose immunosuppression. This necessarily implies that the associated risks such as infections and development of malignancies cannot be avoided.

Future Directions

Two basic problems involved in organ transplantation are the shortage of organs and the side-effects of prolonged non-specific immunosuppression. Xenotransplantation, or transplant of organs from other species, is an exciting possibility, which is being extensively investigated. However, many problems involved such as the transmission of unknown animal viruses and response to foreign xenoantigens, poses major ethical issues

Efforts are also on to develop ideal immunosuppressive drugs since insufficient immunosuppression may cause excessive rejection episodes and excessive immunosuppression results in more opportunistic infections. There is a need for the development of therapeutic modalities that induce graft specific nonresponsiveness without suppressing immune responses to nondonor antigens, including infectious agents and tumour antigens.

Transplant tolerance

The ideal scenario for any transplantation would be to achieve complete tolerance by the recipient for the transplanted organ. However, this type of total donor specific tolerance is not easy to achieve. Pre-transplant transfusions of donor blood to the recipient (Donor specific transfusions) have been used in renal transplantation. The rationale is that repeated exposure to donor HLA antigens present on white blood cells would result in the development of tolerance. However, this strategy has shown limited utility in clinical practice. Recently, the transfusion of small amounts of donor bone marrow into the recipient prior to transplantation of solid organ has provided donor specific tolerance.

Anther approach that has been tried is treatment with antibodies to co-stimulatory antigen CD28, since in the absence of co-stimulation T cells develop anergy. More recently, there is considerable interest in a subpopulation of CD4+ cells termed regulatory T cells, which bear the CD25 cell surface marker and a specific transcription marker, FoxP3. Though the role of these cells in bone marrow transplantation is being extensively studied, their role in solid organ transplantation is less well understood.

Clinical Organ Transplantation

With the understanding of the various issues in transplantation, it should now be clear as to why it is important to select donors with the least genetic disparity from the recipient.

Donor Selection

In all transplantation, the objective is to minimise the graft immunogenicity. To avoid hyper acute rejection, the ABO blood group antigens should always be compatible between donor and recipient. Organs are harvested from cadavers and precious organs such as heart, lung, intestine, liver are grafted to waiting recipients without HLA matching, since these organs cannot be artificially kept functional like the kidney and transplantation becomes an emergency requirement. The availability of these organs is also very limited.

The scenario is different for bone marrow and kidney transplants. In the case of renal transplantation, patients can be maintained in good health for long periods of time on dialysis. Kidneys are grafted both from cadavers and live donors. Donation from live donors is possible as each individual has 2 kidneys and can function normally with one healthy kidney. Although there is a view that this is unethical living donors are the only way by which the large demand for kidneys can be partially met. Kidney donors are selected on the basis of HLA -DR, -A and -B loci matching, these being the most important loci for alloreactivity. In the West, where most renal grafts are taken from cadaver donors, each kidney of the donor is given to 2 recipients with the closest MHC match from among the waiting list of recipients.

Kidneys are flown across US and between continents in order to get the best possible donor recipient combination and thus get maximum benefit of the available organ. In countries like ours, where majority of transplantations are performed from live related donor, efforts are made to find an HLA-identical sibling. In case this is not possible, parent or haplo-identical sib is selected as the donor. Various testing principle, test design, available methodologies and clinical value of test results is shown in **Table 10.3.**

Table 10.3 Various test principle, test design, available methodologies and test results

Testing principle	Test Design	Available methodologies	Clinical value of test results
Determine donor-recipient compatibility	Compare ABO and HLA types of patients and donors	• Serology • Molecular typing	• HLA matching reduce rejection and GVHD
Analyze allosensitization status of patient	Serum screening against HLA-typed panels: • Determine PRA • Define HLA antibody specificity patterns	• Direct and AHG lymphocytotoxicity • ELISA	• Identify high PRA patients • Define acceptable HLA mismatched antigens • Interpret crossmatch results with potential donors
Perform cross match	Test patient serum against potential donors	• Direct and AHG lymphocyto-toxicity • Flow cytometry • ELISA	• Prevent hyper acute rejection • Reduce antibody mediated vascular rejection
Determine T-cell allo-responsiveness between donor and recipient	Mixed leukocyte culture assay	• Proliferation • Cell-mediated cytotoxicity • Cytokine production • Alloreactive T-cell quantitation	• Prognosticators of cellular rejection • Not routinely used in a clinical setting

HLA PCA AHG GVHD

Numerous factors influence the probability of success in solid organ transplantation. Both patient and graft survival can be influenced by factors such as race, histocompatibility, immune responsiveness, age, sex, primary disease, cause of donor death, and other characteristics of the candidate and donor. Factors related to the mechanisms of the allocation, such as kidney ischemic time, and factors related to geography, such as the identity of the transplant centre, also influence outcome.

The scenario in bone marrow transplantation is different. The indications for transplant vary widely from congenital immunodeficiencies and thalssemia to various lymphoid and myeloid malignancies. In some situations, such as thalassemia, the patient can be maintained for long periods of time with transfusions. In the case of acute leukemias the situation demands more urgent grafting. Bone marrow cannot be harvested from cadavers and so live donors are the only sources available. Since the objective is to reconstitute stem cells from the donor, the donor and recipient have to be completely HLA matched, in order to minimise the dangers of GVHD. Conventionally, in BMT, bone is harvested from the iliac crest and passed through a filter to remove bone chips etc. and transfused to the recipient through a peripheral vein. Since the objective is to transfuse the stem cells, stem cells are mobilized from the marrow to the peripheral blood in the donor by treatment with growth factors like granulocyte colony stimulating factor. Following this, donor blood is passed though a cell separator, through which all the blood flows. The white cells are retained by a column and the rest of the components are transfused back. The harvested cells are then transfused to the recipient. This procedure, called peripheral stem cell transplantation or PSCT, has almost completely replaced the conventional approach. It has made the donor more comfortable and the whole procedure less tedious. However, all the problems attached to transfusion of allergenic cells that have been discussed before apply equally to PSCT.

Glossary

Accessory cell: A cell required for, but not actually mediating, a specific immune response.

Acute-phase proteins: Proteins, that participates in the early phases of host defense against infection and show a rapid increase in the blood levels after the onset of an infection or inflammation.

Adaptive immune response: The response of antigen-specific lymphocytes to antigen including the development of immunologic memory; also known as acquired immune response.

Affinity: A measure of the binding constant of a single antigen-combining site with a monovalent antigenic determinant.

Agglutination: The aggregation of particulate antigen by antibodies. Agglutination applies to red blood cells as well as to bacteria and inert particles coated with antigen.

Allele: An alternate form of a gene that occupy the same position or locus on a specific chromosome.

Allelic exclusion: The ability of heterozygous lymphoid cells to produce only one allelic form of antigen-specific receptor when they have the genetic endowment to produce both. Genes other than those for the specific receptors are usually expressed codominantly.

Allergen: An antigen responsible for producing allergic reactions by inducing IgE formation.

Allergy: An immune reaction to non-pathogenic antigens, which lead to inflammation and deleterious effects in the host.

Allogenic: Having a genetic dissimilarity within the same species.

Allograft: A tissue transplant (graft) between two genetically-nonidentical members of a species.

Allotypes: Antigenic determinants that is present in allelic (alternate) forms. When used in association with immunoglobulin, allotypes describe allelic variants of immunoglobulin detected by antibodies raised between members of the same species.

Alternate complement (Alternative) pathway: The mechanism of complement activation that does not involve activation of the C1, C4, C2 pathway by antigen-antibody complexes, and begins with the activation of C3 by factor B.

Anaphylaxis: Immediate hypersensitivity response to antigenic challenge, mediated by IgE and mast cells. It is a life-threatening allergic reaction, caused by the release of pharmacologically active agents.

Anergy: A state of antigen-specific non-responsiveness in which antigen specific T or B cells are present but functionally unable to respond to antigen.

Antibody: A serum protein formed in response to foreign antigen; antibodies are generally defined in terms of their specific binding to the immunizing antigen.

Antibody-dependent, cell-mediated cytotoxicity (ADCC): A phenomenon in which target cells, coated with antibody, are destroyed by specialized killer cells (NK cells and macrophages), which bear receptors for the Fc portion of the coating antibody (Fc receptors). These receptors allow the killer cells to bind to the antibody-coated target.

Antigen presentation: The display of antigen as peptide fragments bound to MHC molecules on the surface of a cell; T cells recognize antigen only when it is presented in this way.

Antigen processing: The degradation of proteins into peptides that can bind to MHC molecules for presentation to T cells. All protein antigens must be processed into peptide before MHC molecules can present them.

Antigen receptor: The specific antigen-binding receptor on T or B-lymphocytes.

Antigen: Any foreign material that is able to trigger an immune response; also used loosely to describe materials used for immunization.

Antigen-binding site: The part of an immunoglobulin molecule that binds antigen specifically.

Antigenic determinant: A single antigenic site or epitope on a complex antigenic molecule or particle.

Antigen-presenting cell (APC): A specialized type of cell, bearing cell surface class II MHC (major histocompatibility complex) molecules, involved in processing and presentation of antigen to inducer, or helper, T cells. Examples: macrophage, dendritic cells.

Antiserum (plural: antisera): Serum containing antibodies against the molecules used for immunization.

Apoptosis: A form of cell death caused by activation of endogenous molecules leading to the fragmentation of DNA.

Atopy: A term used to describe IgE-mediated anaphylactic responses in humans; usually genetically determined.

Autograft: A tissue transplant from one area to another on the same individual.

Autoimmunity (autoallergy): An immune response to "self" tissues or components. Such an immune response may have pathological consequences leading to autoimmune diseases.

Autosome: A non-sex-determining chromosome. Cells have 22 pairs of autosomes and a pair of sex-determining chromosomes.

Avidity: The strength of antigen to antibody binding when multiple epitopes on an antigen interact with multiple binding sites of an antibody; it is the summation of multiple affinities.

B lymphocyte (B cell): The precursors of antibody-forming plasma cells; these cells carry immunoglobulin and class II MHC (major histocompatibility complex) antigens on their surfaces.

BCR: The complex of membrane bound immunoglobulin along with the Igα and Igβ chains on B-lymphocytes.

Base Pair: (bp) Two complementary-building blocks contained in DNA. The nucleotides adenine and thymine bond to make a base pair in DNA, cytosine and guanine form another pair. Approximately 150 million base pairs make up each human chromosome.

Basophil: A polymorphonuclear leukocyte with blue-staining granules, which contain heparin, histamine and other vasoactive amines. Within tissues, these cells are known as mast cells

Bence-Jones protein: Immunoglobulin light chains in the urine of patients with multiple myeloma.

Blocking antibody: A functional term for an antibody molecule capable of blocking the interaction of antigen with other antibodies or with cells.

Bone marrow: The site of hematopoiesis, in which stem cells give rise to the cellular elements of blood, including red blood cells, monocytes, polymorphonuclear leukocytes, platelets, and lymphocytes.

Bursa of Fabricius: An out pouching of the cloacae in birds; site of development of B cells in birds.

Carcinoma embryonic antigen (CEA): A protein expressed transiently during embryonic development; gets re-repressed in malignant tissue.

Carrier: A person who has both a recessive mutated gene and a normal form of the same gene. Carriers do not usually develop the disease associated with the gene mutation, but they can pass the mutation on to their children.

Cell-mediated cytotoxicity (CMC): Killing (lysis) of a target cell by an effector lymphocyte.

Cell-mediated immunity (CMI): Immune reaction mediated by T cells; in contrast to humoral immunity, which is antibody mediated. Also referred to as delayed-type hypersensitivity.

Chemotaxis: Migration of cells along a concentration gradient of chemoattractant.

Chromosome: A structure in the nucleus of a cell that contains genes. Normal human cells have 46 chromosomes, which are arranged in pairs.

Class I, II and III MHC molecules: Proteins encoded by genes in the major histocompatibility complex. Class I molecules are HLA-A, B, or C. Class II molecules are designated DP, DQ or DR.

Classical complement pathway: The mechanism of complement activation initiated by antigen-antibody aggregates and involving generation of C4b C2a.

Clonal deletion: The loss of lymphocytes of a particular specificity due to contact with either "self" antigen.

Clonal selection theory: The prevalent concept that specificity and diversity of an immune response are the result of selection by antigen of specifically reactive clones from a large repertoire of preformed lymphocytes, each with individual specificities.

Clone: A group of identical genes, cells or organisms that come from a single ancestor.

Cloning: The process of making genetically-identical copies.

Cluster differentiation (CD): A collection of monoclonal antibodies that recognize antigen found on a particular differentiated cell type or types. Each of the antigens recognized by such a collection of antibodies is called a CD markers.

Combinatorial joining: The joining of segments of DNA to generate essentially new genetic information, as occurs with Ig genes during the development of B cells. Combinatorial joining allows multiple opportunities for 2 sets of genes to combine in different ways.

Complement receptor: Membrane anchored proteins on erythrocytes, lymphocytes, neutrophils, monocytes and macrophages that binds to be fragments of C3.

Constant region (C region): The invariant carboxyl-terminal portion of an antibody T cell receptor molecule.

Cross-reactivity: The ability of an antibody, specific for one antigen, to react with a second antigen; a measure of relatedness between two different antigenic substances.

Cytokines: Soluble substances secreted by cells, which have a variety of effects on other cells.

Cytotoxic (Cytolytic) T cell: A T-cell that can kill target cells bearing the appropriate antigen within the groove of an MHC class I molecule which is recognized by its TCR.

Delayed type hypersensitivity (DTH): A T cell-mediated reaction to antigen, which takes 24-48 hours to develop, and involves release of lymphokines and recruitment of monocytes and macrophages.

Determinant: Part of the antigen molecule, which binds to an antibody-combining site or to a receptor on T cells (see hapten and epitope).

Differentiation antigen: A cell surface antigenic determinant found only on cells of a certain lineage and at a particular developmental stage; used as an immunologic marker.

DNA (Deoxyribonucleic Acid): A large molecule with the complete set of instructions for making all the proteins a cell needs. DNA consists of two spiral strands of millions of chemical building blocks or base pairs.

Domain: A compact segment of an immunoglobulin molecule, made up of about 110 amino acids around an S-S bond, and encoded by a unique segment of DNA, surrounded by non-translated sequences.

Dominant allele: A form or mutation of a gene that will produce a trait or disease if a copy is inherited from just one parent.

Enhancing antibodies: Antibodies which enhance the survival of a graft or of a tumor.

Eosinophil chemotactic factor of anaphylaxis (ECF-A): A substrate released from mast cells during anaphylaxis, which attracts eosinophils.

Eosinophil: A polymorphonuclear leukocyte with large eosinophilic (i.e. red) cytoplasmic granules.

Epitope: An alternative term for antigenic determinant.

Equivalence zone: The region of equivalence of antigen and antibody, which results in precipitation of antigen – antibody complex.

Exon: The region of DNA coding for a protein or a segment of a protein.

F(ab'): A fragment generated by cleavage an Ig antibody molecule with the enzyme pepsin which contains two antigen-combining sites.

Fab: Fragment of Ig containing one antigen-binding site, generated by cleavage of the antibody with the enzyme papain. This enzyme, which cuts at the hinge region N-terminal to the inter-H-chain disulphide bond and generates two Fab fragments from one antibody molecule.

Fc receptor (FcR): A receptor on a cell surface with specific binding affinity for the Fc portion of an antibody molecule. Fc receptors are found on many types of cells.

Fc: Fragment of Ig molecule containing the C-terminal domains of the heavy immunoglobulin chains generated by cleavage with papain.

Freund's complete adjuvant: A water-in-oil emulsion that contains an emulsifying agent, and killed mycobacterium which enhances the immune response to the immunogen; termed "incomplete" Freund's adjuvant if mycobacterium is not included.

Gene: A piece of DNA containing instructions for making a specific protein. The body has approximately 30,000 genes, each with a different role in heredity. For example, the protein developed by one gene may determine hair color, and another, when defective, may cause a disease.

Gene conversion: A process often associated with recombination, during which one allele is replicated at the expense of another, leading to non-mendelian segregation ratios. In whole tetrads, for example, the ratio may be 6:2 or 5:3 instead of the expected 4:4.

Genetics: The study of heredity, focusing on how particular qualities or traits are transmitted from parents to children.

Genome: The total genetic material in the chromosomes of an organism.

Genotype: A person's genetic makeup for one specific trait.

Germ line: Refers to genes in germ cells as opposed to somatic cells.

Graft versus host reaction (GVH): The pathologic consequences of a response initiated by transplanted immunocompetent T lymphocytes into an allogenic, immunologically incompetent host.

H-2 complex: The major histocompatibility complex situated on chromosome 17 of the mouse; contains sub-regions K, D and I.

Haplotype: A particular combination of closely-linked genes on a chromosome inherited from one parent.

Hapten: A compound, usually of low molecular weight, that is not itself immunogenic but that, after conjugation to a carrier protein or cells, becomes immunogenic and induces antibody, which can bind the hapten alone in the absence of carrier.

Heavy chain (H chain): The larger of the two types of chains that comprise a normal immunoglobulin or antibody molecule.

Helper T cells: A class of T cells which help to trigger B cells to make antibody against thymus-dependent antigens. Helper T cells also help to generate cytotoxic T cells.

Heterophile antigen: A cross-reacting antigen that appears in widely ranging species such as humans and bacteria.

Hinge region: A flexible, open segment of an antibody molecule that allows bending of the molecule. The hinge region is located between Fab and Fc and is susceptible to enzymatic cleavage.

Human Genome Project: An international effort to sequence the total DNA on the 46 human chromosomes.

Humoral immunity: Any immune reaction that can be transferred with immune serum is termed humoral immunity (as opposed to cell-mediated immunity).

Hybrid: An organism with different genes for the same trait.

Hybridoma: Immortalized cell line that secretes antibody of one defined antigen specifically.

Hypersensitivity: State of reactivity to antigen that is greater than normal for the antigenic challenge; hypersensitivity is the same as allergy and denotes a deleterious outcome rather than a protective one.

Hyper variable regions: Portions of the light and heavy immunoglobulin chains that are highly variable in amino acid sequence from one immunoglobulin molecule to another, and that, together, constitute the antigen-binding site of an antibody molecule. Also portions of the T-cell receptor, which constitute the antigen-binding site.

Idiotype: The combined antigenic determinants (idiotopes) found on antibodies of an individual that are directed at a particular antigen;

such antigenic determinants are found only in the variable region.

Immediate-type hypersensitivity: Hypersensitivity tissue reaction occurring within minutes after the interaction of antigen and antibody.

Immune adherence: The adherence of particulate antigen coated with C3b to tissue having cells with C3b receptors.

Immune complex: A complex of antigen and antibody molecules.

Immune modulators: Substances that alter the control the expression of the immune response.

Immune response (Ir) gene: A gene controlling an immune response to a particular antigen; most genes of this type are in the MHC (major histocompatibility complex), and the term is rarely used to describe genes outside the MHC.

Immunogen: A substance capable of inducing an immune response (as well as reacting with the products of an immune response).

Immunoglobulin (Ig): A general term for all antibody molecules. Each Ig unit is made up of two heavy chains and two light chains and has two antigen- binding sites.

Interferon: A group of proteins having antiviral activity and are capable of enhancing and modifying the immune response.

Interleukins: Glycoproteins secreted by a variety of leukocytes which have effects on other leukocytes.

Internal image: A spatial configuration of the combining site of an anti-idiotype antibody which resembles the epitope to which the idiotype is directed.

Intron: Segment of DNA that exist in the intervening sequence of nucleotides between coding sequences or exons.

Isograft: A tissue transplanted between two genetically-identical individuals.

Isohemagglutinins: Antibodies to major red blood cell antigens present normally as a result of inapparent immunization by cross-reactive antigens in bacteria, food, etc.

Isotype switch: The shift of a B cell or its progeny from the secretion of antibody of one isotype or class of antibody with the same V regions but a different heavy- chain constant region and, hence, a different isotype (class switch).

Isotypes: Classes of antibody that differ in the constant region of their heavy chain (Fc portion). They have different biological activities.

J chain (joining chain): A polypeptide involved in the polymerization of immunoglobulin molecules IgM and IgA.

K cell: An effector lymphocyte with Fc receptors, which allow it to bind to and kill antibody-coated target cells.

Light chain (L chain): An immunoglobulin chain either kappa or lambda, which combines with H chain to form Ig molecules.

Lymphocyte: Small cell with scanty cytoplasm, found in blood, in all tissue, and in lymphoid organs, such as lymph nodes, spleen, and Peyer's patches, and bears antigen-specific receptors.

Lymphokines: Soluble substances secreted by lymphocytes, which have a variety of effects on lymphocytes and other cell types.

Macrophage: A large phagocytic cell of the mononuclear series found within tissues. Properties include phagocytosis, and antigen presentation to T cells.

Macrophage-activating factor (MAF): A group of several lymphokines, including interferon, released by activated T cells, which together induce activation of macrophages, making them more efficient in phagocytosis and cytotoxicity.

Major histocompatibility complex (MHC): A cluster of genes on chromosome 6 in humans, encoding cell surface molecules important for presentation of peptides to TCR. Several classes of protein such as MHC class I and II proteins are encoded in this region. In humans, these are known as 'Human leukocyte antigens' (HLA).

Mast cell: Tissue located cell, probably derived from basophils. Possesses receptor for Fc of IgE. Participates in 'Immediate hypersensitivity' reactions.

MHC class I molecule: A molecule encoded by genes of the MHC which participates in antigen presentation to cytotoxic T (CD8+) cells.

MHC class II molecule: A molecule encoded by genes of the MHC which participates in antigen presentation to helper T (CD4+) cells.

MHC restriction: The ability of T lymphocytes to respond only when they 'see' the appropriate antigen in association with "self" MHC class I or class II proteins on the antigen presenting cells.

Migration inhibition factor (MIF): A lymphokine that inhibits the motility of macrophages in culture.

Minor histocompatibility antigens: These antigens, encoded outside the MHC, are numerous, but do not generate rapid graft rejection or primary responses of T cells in vitro. They do not serve as restricting elements in cell interactions.

Mitogen: A substance that stimulates the proliferation of many different clones of lymphocytes.

Mitosis: The process of self-replication by a cell. During this stage of a normal cell's life cycle, the parent cell divides into two identical cells.

Mixed lymphocyte reaction (MLR): When lymphocytes from two individuals are cultured together, a proliferative response is generally observed, as the result of reactions of T cells of one individual to MHC antigens on the other individual's cells.

Molecule: The smallest physical unit of a chemical substance.

Monoclonal: Product of a single clone. A clone is the progeny of a single cell. In immunology, monoclonal generally describes a preparation of antibody that is specific to a single antigen specificity.

Monocyte: Large circulating white cell, 2-10% of total white cells, phagocytic, indented nucleus. Migrates to tissues, where it is known as a macrophage.

Monokines: Soluble substances secreted by monocytes, which have a variety of effects on other cells.

Mutation: A change in genetic material, such as the number, arrangement, or molecular sequence of a gene.

Myeloma: A tumor of plasma cells, generally secreting a single species of immunoglobulin.

NK cell: Naturally occurring, large, granular, lymphocyte-like cells that kill various tumour cells; they may play a role in resistance to tumors. Also, they participate in ADCC. They do not exhibit antigen specificity, and their number does not increase by immunization.

Null cells: An early population of lymphocytes bearing neither T-cell nor B-cell differentiation antigens.

Opsonin: A substance, usually antibody or complement component, which coats a particle such as a bacterium and enhances phagocytosis by phagocytic cells.

Opsonization: The coating of an antigen with antibody and/or complement that leads to enhanced phagocytosis of the antigen by phagocytic cells.

Paratope: An antibody combining site that is complementary to an epitope.

Passive immunization: Immunization by the administration of preformed antibody into a nonimmune individual.

PCR (Polymerase Chain Reaction): A technique used to make millions of copies of DNA from a tiny sample.

Penetrance: The probability that a particular gene mutation will produce the disease.

Phagocytosis: The engulfment of a particle or a microorganism by leukocytes.

Phenotype: The observable expression of a genetically determined trait. The trait may be visibly apparent, such as hair color, or detectable with a laboratory test, such as blood type.

Pinocytosis: Ingestion of liquid or very small particles by vesicle formation in a cell.

Plasma cell: End-stage differentiation of a B cell to an antibody-producing cell.

Polyclonal activator: A substance that induces activation of many individual clones of either T or B cells.

Polymorphonuclear leukocyte: White cell, granular cytoplasm, phagocytic, neutral staining (neutrophil). Basophilic staining (basophil) or eosinophilic staining (eosinophil).

Primary lymphoid organs: Organs in which the maturation of T and B lymphocytes takes place and antigen-specific receptors are first acquired.

Primary responses: The immune response to a first encounter with antigen. The primary responses are generally small, have a long induction phase or lag period, consist primarily of IgM antibodies, and generates immunologic memory.

Probe: A length of single-stranded DNA that matches part of a gene that has already been identified. If the DNA probe is made from a mutated gene, it can be used to determine whether someone has that genetic defect. The probe will bind to complementary parts of the patient's gene.

Proteins: A large, complex molecule made of amino acids. Hormones, enzymes, and antibodies are examples of proteins.

Radioallergosorbent test (RAST): A solid-phase radioimmunoassay for detecting IgE antibody specific for a particular allergen.

Radioimmunoassay (RIA): A widely used technique for measurement of primary antigen-antibody interactions, and for the determination of the level of important biological substances in mixed samples. It takes advantage of the specificity of the antigen-antibody interaction and the sensitivity that derives from the measurement of radioactively labeled materials.

Reagin: Term used in allergy for IgE antibodies.

Recessive allele: A form or mutation of a gene that must be inherited from both parents to produce a trait or disease.

Respiratory burst: Oxygen dependent increase in metabolic activity within phagocytic cells stimulated by bacteria or parasites, to be microbicidal.

Reticulo endothelial system: A network of phagocytic cells.

Rheumatoid factor: An autoantibody (usually IgM) which reacts with the individual's own IgG. Present in rheumatoid arthritis.

RNA (Ribonucleic Acid): A chemical similar to DNA from which proteins are made. Unlike DNA, RNA can leave the nucleus of the cell.

Second set rejection: Accelerated rejection of an allograft in an already immune recipient.

Secondary lymphoid organs: Organs in which antigen-driven proliferation and differentiation of B and T lymphocytes takes place.

Secretory component: A surface receptor on epithelial cells lining mucosal surfaces which binds dimeric IgA and transports it through the cell into mucosal secretions.

Serum sickness: A hypersensitivity reaction consisting of fever, rashes, joint pain and glomerulonephritis, resulting from localization of circulating, soluble, antigen-antibody complexes, which induce inflammatory reactions.

Slow-reacting substance of anaphylaxis (SRS-A): A group of substances released by mast cells during anaphylaxis which induces a prolonged constriction of smooth muscle. This prolonged constriction is not reversible by treatment with antihistamines.

Stem Cells: Unspecialized, immature cells in the bone marrow from which all types of blood cells originate.

T cell: A lymphocyte which undergoes a developmental stage in the thymus.

T-dependent antigen: An immunogen that is able to induce antibody synthesis only in the presence of lymphokines released by helper T cells.

T-independent antigen: An immunogen which induces antibody synthesis in the absence of lymphokines released by T cells; the antibodies are generally only of the IgM isotype.

Tolerance: Diminished or absent capacity to make a specific response to an antigen, usually produced as a result of contact with that antigen under nonimmunizing conditions.

Toxoid: A non-toxic derivative of a toxin used as an immunogen for the induction of antibodies capable of cross-reacting with the toxin.

Trait: A physical or temperamental characteristic, such as eye color, leg length etc.

Transcription: The process of copying information from DNA into new strands of messenger RNA (mRNA), which then carries the information to the cytoplasm to serve as a blueprint for manufacturing a specific protein.

Tumor-Suppressor Genes: Genes that normally restrain cell growth. If they are defective or missing, cancer can develop because cells can grow uncontrollably.

Vaccination: Originally referred to immunization against smallpox with the less virulent cowpox (vaccinia) virus; more loosely used for any immunization against a pathogen.

Vector: A "vehicle", such as a modified virus or DNA molecule, used to deliver genetic material into the body for gene therapy.

Index